Evolution of Matter and Energy

M. Taube

Evolution of Matter and Energy
on a Cosmic and Planetary Scale

With 136 Illustrations

Springer-Verlag
New York Berlin Heidelberg Tokyo

M. Taube
Eidgenössisches Institut für Reaktorforschung
Institut Fédéral de Recherches en Matière de Réacteurs
CH-5303 Würenlingen
Switzerland

Cover photo courtesy of NASA.

Library of Congress Cataloging in Publication Data
Taube, M. (Mieczyslaw), 1918–
  Evolution of matter and energy on a cosmic and
planetary scale.
  Bibliography: p.
  Includes index.
  1. Matter. 2. Force and energy. 3. Cosmology.
I. Title.
QC171.2.T37   1985   530   84-26877

© 1985 by Springer-Verlag New York Inc.
All rights reserved. No part of this book may be translated or reproduced in any
form without written permission from Springer-Verlag, 175 Fifth Avenue,
New York, New York 10010, U.S.A.

Typeset, printed, and bound by Graphischer Betrieb Konrad Triltsch, Würzburg,
Federal Republic of Germany.
Printed in the Federal Republic of Germany.

9 8 7 6 5 4 3 2 1

ISBN 0-387-13399-2 Springer-Verlag New York Berlin Heidelberg Tokyo
ISBN 3-540-13399-2 Springer-Verlag Berlin Heidelberg New York Tokyo

Preface

My intention in this book is to describe in simple language, using a minimum of mathematics but a maximum of numerical values, the most important developments of science dealing with matter and energy on cosmic and global scales. In the conventional literature all of these findings are distributed among books and journals on physics, astronomy, chemistry, geology, biology, energy, engineering, and the environmental sciences.

The main purpose here is to attempt to give a unified description of Nature from the elementary particles to the Universe as a whole. This is used as a basis for analysing the future development of mankind.

The future evolution of the Universe, galaxies, stars, and planets gives some hope for the destiny of mankind. The problem of matter and energy flow on the Earth appears soluble even for the distant future. There seems to be no reason why a long period of human development on this planet should not be possible.

The book has been prepared based on my lectures at the Warsaw University from 1959 to 1968 and during the 15 years 1969–1983 at the Swiss Federal Institute of Technology (Eidgenössische Technische Hochschule) in Zürich and at the University of Zürich.

I wish to give my sincere thanks to the Swiss Federal Institute for Reactor Research at Würenlingen for their constant support.

I am especially grateful to Mrs. Christine Stratton for setting up the English text and to Mr. R.W. Stratton and I.G. McKinley for their helpful criticisms and remarks.

Finally, I would like to acknowledge the excellent work of Mrs. T. Christen in preparing the typescript.

Killwangen, Kanton Aargau, Switzerland   M. Taube
1984

Contents

CHAPTER 1
Matter and Energy. The Interplay of Elementary Particles and Elementary Forces

1.1 An Attempt to Describe the Natural World Using the Smallest Number of Elementary Phenomena.	1
1.2 General Foundations of the Physical Sciences	4
1.2.1 Some principles	4
1.2.2 Some properties of the elementary phenomena are governed by very exact and strong laws of conservation.	4
1.2.3 Prohibitions	5
1.3 Elementary Forces and Particles	5
1.3.1 Elementary forces	5
1.3.2 Elementary particles	7
1.4 Elementary Particles	8
1.4.1 "Bricks" and "mortar".	8
1.4.2 Creation of the elementary particles.	9
1.4.3 "Life" and "death" of elementary particles.	11
1.5 The Existence of Atomic Nuclei Is Due to the Forces of Attraction Between Their Nucleons.	12
1.5.1 The weak force limits the number of stable hadrons.	12
1.5.2 Strong force binds the nucleons together	13
1.5.3 Binding energy of a nucleon	14
1.6 Matter and Free Energy – The Intimate Connection	15
1.7 What Are the Conclusions for the Future Development of Mankind?.	17

CHAPTER 2
The Universe: How Is It Observed Here and Now? Its Past and Possible Future

2.1 What Is the Universe?.	19
2.1.1 A definition of the Universe	19
2.1.2 Beginning of the Universe	19

2.2 Expansion of the Universe. 21
   2.2.1 The red shift. 21
   2.2.2 The five eras of the Universe. 21
2.3 What Is Known About the Universe Today? 21
   2.3.1 The average composition of the Universe. 21
   2.3.2 Chemical composition of cosmic matter. 24
   2.3.3 Composition of photons 25
2.4 The Universe as a Whole 26
2.5 The Future of the Universe 28
2.6 What Conclusions Can Be Drawn for the Future Development
   of Mankind? 29

CHAPTER 3

The Origin and Nuclear Evolution of Matter

3.1 The Creation of the Elementary Particles in the Very Early Universe . . . 30
   3.1.1 Unknown phase: Era of superunified force (Planckian Era or
         Very Hot Era). 30
   3.1.2 Era of grand unified force (Hot Era) 31
   3.1.3 Era of unified force (Lukewarm Era) 31
   3.1.4 Cold Era and Very Cold Era. 32
3.2 Evolution of the Elementary Particles. A Very Rapid Development
   in the First Seconds of the Universe 33
   3.2.1 Beginning of the Cold Era: Evolution in the "Hadron Epoch". . . 33
   3.2.2 Production of hydrogen, deuterium, and helium: The Universe a
         few seconds old; Lepton Epoch 33
   3.2.3 The Photon Epoch, from the first minute to the first million years. 36
3.3 The Beginning of the Present Very Cold Era: The "Stars Era".
   The Evolution of Galaxies, Stars, and Life. 37
   3.3.1 The largest of the cosmic structures: The development of galaxies . 37
   3.3.2 The evolution of stars; the nuclear and gravitational reactors . . . 37
   3.3.3 The protostar evolves from diffuse matter 39
   3.3.4 The longest living stars, those of the Main Sequence 42
   3.3.5 Red Giants: The cold stars with the hot interiors 43
   3.3.6 Evolution towards hot dense stars. 44
   3.3.7 Explosion of a supernova: The most spectacular event in a galaxy. 45
   3.3.8 Extremely dense stars: Neutron stars (pulsars) and black holes. . . 45
3.4 The Burning of Hydrogen – Nucleosynthesis in the Stars. 46
   3.4.1 Deuterium: The fuel of protostars. 46
   3.4.2 The slow burning of hydrogen. 47
   3.4.3 The burning of hydrogen in a catalytic cycle assisted by carbon,
         nitrogen, and oxygen 49
3.5 Helium also Burns, but under More Extreme Conditions. 49
   3.5.1 Production of carbon from the burning of helium 49
   3.5.2 A very vital step: The production of oxygen 50

3.6 Carbon, Oxygen, and Other Elements of Medium Mass Burn in a Flash . 50
   3.6.1 Energy production and energy required for nucleosynthesis. 50
   3.6.2 Iron, the nuclear ash 50
3.7 The Systhesis of Heavy Elements: The Need for an External Energy Source 51
   3.7.1 How can uranium be synthesised?. 51
   3.7.2 The "s-process", the slow-process of neutron capture 52
   3.7.3 The "r-process", the rapid-process of neutron capture. 53
3.8 Cosmic Rays – A Strange Form of Matter. 55
3.9 What Are the Conclusions for the Future of Mankind? 56

CHAPTER 4

Chemical Evolution and the Evolution of Life: The Cosmic Phenomena

4.1 Chemical Evolution: Another Phase in the Evolution of Matter 58
   4.1.1 Special case of the electromagnetic force: The chemical force. . . . 58
   4.1.2 The actors in the chemical play 58
4.2 Chemical Synthesis Occurs in Cosmic Space. 62
   4.2.1 Interstellar gas contains very many, often very complex compounds 62
   4.2.2 Some of the interstellar molecules exist in solid form 63
   4.2.3 Comets: Rare and strange, but formidable, chemical reactors. . . . 65
   4.2.4 Meteorites often consist of very "sophisticated" chemical compounds 66
   4.2.5 "Organic molecules" on the Moon and planets 67
4.3 The Origin of the Planets 70
   4.3.1 Have the planets been formed "by chance"? 70
   4.3.2 The protoplanet, the first stage of evolution 71
   4.3.3 The chemical evolution of the Earth: A complex and dramatic development. 72
   4.3.4 All stable elements present in the Universe exist on Earth. 75
   4.3.5 The history of the Earth has been influenced by the movement of the continents 77
   4.3.6 The first phases of chemical evolution were driven by different energy sources and were influenced by a number of factors. 78
4.4 Synthesis of Complex Molecules on the Primitive Earth 82
   4.4.1 The primitive atmosphere includes mostly molecules containing hydrogen. 82
   4.4.2 The amino acids; their ease of synthesis 83
   4.4.3 How were the large molecules, the polymers, produced? 84
4.5 What Is Life? The Need for a General Definition. 85
   4.5.1 Could life have originated spontaneously? 85
   4.5.2 The physical aspect of life 86
   4.5.3 What kind of elementary forces can play the role of energy carriers for living systems? 88
   4.5.4 What kind of elementary particles can play the role of carriers of life? 88

- 4.6 The Chemical Elements, Particularly the Light Elements, Are the Carriers of Life 89
  - 4.6.1 Why are the light elements best fitted for this role? 89
  - 4.6.2 Why is hydrogen oxide – water – the unique medium for living organisms?. 91
  - 4.6.3 The source of free energy for life: The stars of the Main Sequence. 92
  - 4.6.4 The chemical composition of the living organism is similar to the chemical composition of the Universe. 94
  - 4.6.5 Life is only possible in a Universe having the characteristics of our type of Universe. 96
- 4.7 What Can We Hope to Know About the Spontaneous Formation of Terrestrial Life? 97
  - 4.7.1 The problem: The uniqueness of life in our present state of knowledge. 97
  - 4.7.2 The protobionts: The first living structures. 98
  - 4.7.3 The evolution of the living being occurred at the switch-over point from one energy source to the next 98
- 4.8 Evolution of Living Beings 101
  - 4.8.1 Genetic evolution. 101
  - 4.8.2 The evolution of Man 102
  - 4.8.3 The evolution of the brain 103
- 4.9 What Are the Conclusions for the Future of Mankind? 104

CHAPTER 5

The Eternal Cycle of Matter on the Earth

- 5.1 Matter on This Planet Is Almost Indestructible 105
  - 5.1.1 How stable is terrestrial matter?. 105
  - 5.1.2 Terrestrial matter is isolated by the gravitational field; the amount of matter is constant 106
  - 5.1.3 Division of the Earth into five "spheres". 106
- 5.2 The Gaseous Sphere Acts in the Exchange Between the Other Spheres 108
  - 5.2.1 The main components of the atmosphere. 108
  - 5.2.2 The most active component, oxygen, a product of the biosphere. 110
  - 5.2.3 Ozone: Modified oxygen which acts as a shield for the biosphere. 112
  - 5.2.4 The carbon cycle, a chain directly related to the flow of energy in the biosphere and technosphere. 113
  - 5.2.5 The "inert" nitrogen cycle, which controls the activity of the biosphere 116
  - 5.2.6 The micro-components of the atmosphere, the troublesome "details" 117
  - 5.2.7 Dust particles, a troublesome constituent of the atmosphere 119
- 5.3 The Hydrosphere – A Crucial Factor in the Existence of the Biosphere. 120
  - 5.3.1 The cycling of water, the largest terrestrial material cycle. 120
  - 5.3.2 Quality of water, quality of life 122

Contents   xi

    5.3.3   Man's demand for water is gigantic 123
    5.3.4   Drinking water, where purity counts 124
    5.3.5   The erosion of the planetary surface 125
5.4 The Solid Earth, the Litosphere 125
    5.4.1   The main components of the Earth's crust 125
    5.4.2   The Earth's crust, the main source of materials for our civilisation 126
    5.4.3   Metals "prepared" by Nature, the most widely used. 128
5.5 Ordered Matter and Entropy 130
    5.5.1   Concentration means increase of order and decrease of entropy . . 130
    5.5.2   Impact of substances in very small amounts: Poisons 131
    5.5.3   Material dissipation and waste formation increases entropy. 132
5.6 What Are the Conclusions for Mankind's Future Development? 133

CHAPTER 6

The Flow of Energy on the Earth

6.1 The Source of Free Energy on the Earth 135
    6.1.1   The quality of energy: The ordered and disordered forms 135
    6.1.2   The elementary forms of energy 136
    6.1.3   How large is flux of energy? 137
6.2 The Energy Sources on the Earth 139
    6.2.1. Solar energy – The most important source 139
    6.2.2   Spectrum and albedo of solar light 140
6.3 Solar Energy and Climate 142
    6.3.1   The solar energy flux is not constant 142
    6.3.2   Solar energy is transformed into numerous forms and types of energy 145
    6.3.3   The past and future of the terrestrial climate 147
    6.3.4   The local climate depends on continental drift 148
6.4 Non-solar Terrestrial Energy Sources 150
    6.4.1   Other non-solar flows of energy play a small but not insignificant role 150
    6.4.2   The importance of the amount of stored energy 151
6.5 How Much Energy Does Man Need? 152
    6.5.1   Does man need energy at all? 152
    6.5.2   The sources of energy are changeable 154
6.6 The Indirect Use of Solar Energy 156
    6.6.1   The biosphere as Man's energy source for technology 156
    6.6.2   Transformation of solar into kinetic energy: Wind 157
    6.6.3   Transformed solar energy: The kinetic energy of falling water . . . 158
    6.6.4   The "insignificant" form of solar energy: The heat of the oceans . 159
    6.6.5   The best forms of stored solar energy: Oil and coal 159
6.7 The Direct Technological Use of Solar Energy 163
    6.7.1   The simplest way: Space heating 163
    6.7.2   Solar energy converted into electricity on the Earth's surface 164
    6.7.3   The extraterrestrial conversion of solar into electrical energy 166

- 6.8 Technological Use of Non-solar (Nuclear) Energy 167
  - 6.8.1 The heaviest elements: The gift of the supernova 167
  - 6.8.2 Geothermal energy results from the nuclear decay of radionuclides 168
  - 6.8.3 The fission of the heavy nuclides is one of the most abundant terrestrial energy sources..................... 169
  - 6.8.4 Fusion: The second coming of nuclear energy 174
- 6.9 Are There Other Sources of Energy?...................... 177
- 6.10 Energy Production as a Source of Dangerous Waste and Environmental Problems... 179
  - 6.10.1 Energy production and nonradioactive waste materials...... 179
  - 6.10.2 Radioactive waste from nuclear energy................. 181
  - 6.10.3 Are fission reactors really dangerous?................. 182
  - 6.10.4 Radioactive waste and its management................. 183
  - 6.10.5 Fusion: The controlled thermonuclear reactor – Is this the "clean" solution?.................................. 185
  - 6.10.6 Thermal waste, the local and global problem 186
  - 6.10.7 Surface waste in the production of energy 186
- 6.11 The Economics of Energy Production...................... 188
  - 6.11.1 The energy cost of energy 188
  - 6.11.2 What is the price of energy?........................ 189
- 6.12 What Are the Conclusions for the Future Development of Mankind?.. 190

CHAPTER 7

The Biosphere: The Coupling of Matter and the Flow of Free Energy

- 7.1 The Biosphere: The Coupling of Matter and the Flow of Free Energy .. 192
- 7.2 The Terrestrial Biosphere: Mass and Productivity............... 193
  - 7.2.1 The greatest component of the biosphere is, in terms of mass, in the form of trees 193
  - 7.2.2 The biosphere's productivity does not match its pattern of distribution 196
  - 7.2.3 The surprisingly simple chemical composition of the biosphere.... 198
- 7.3 The Magnitude of the Flow of Energy in the Biomass 199
  - 7.3.1 The direct net flux of energy in the biosphere is some 92 TW 199
  - 7.3.2 The total solar energy flux consumed by the biosphere 200
  - 7.3.3 The biosphere in the past............................. 201
  - 7.3.4 The green plant is not only a synthesiser, it is also a water vapouriser 204
- 7.4 The Biosphere as a Source of Food for Mankind................ 206
  - 7.4.1 How much free energy in the form of food does Man need?..... 206
  - 7.4.2 Man requires numerous structural materials for his body....... 206
  - 7.4.3 The winning of food from the biosphere 208
- 7.5 Agriculture, Source of Food for Humans...................... 209
  - 7.5.1 Agricultural requirements of the average man 209
  - 7.5.2 Human food quality 211

7.6 Constraints on the Further Development of Agricultural Production 213
    7.6.1 Can the area under cultivation be increased?. 213
    7.6.2 Can agricultural production be doubled over the next 50 years?. . . 215
7.7 The Ocean. A Source of Human Food? 216
    7.7.1 How productive is the ocean? 216
    7.7.2 The ocean is an important source of proteins 219
7.8 Food Production Needs a Large Energy Input 220
    7.8.1 Solar and technological energy input to agriculture 220
    7.8.2 Single-Cell protein – A new food source 222
7.9 The Biosphere Is More than a Source of Food 223
7.10 What Conclusions Can Be Drawn for Mankind's Future Development? . 223

CHAPTER 8

Is the Future Development of Mankind on This Planet Possible?

8.1 Is It Possible to Consider the Future? 226
8.2 The Main Problem: The Increase of the World Population 226
    8.2.1 Is it wrong to consider mankind as part of the biosphere? 226
    8.2.2 The growth of world population in the past 227
    8.2.3 The reference case used in this chapter – A stable world population of 8 billion. 228
8.3 Problem No. 2: A Place on the Earth for Everyone. 229
    8.3.1 How much space will each inhabitant have in the future?. 229
    8.3.2 Organisation of space and transport: The energy lost 231
8.4 Problem No. 3: Food for Everyone. 231
8.5 Problem No. 4: Material Resources for Everyone 233
    8.5.1 Maximum recycling and minimum use 233
    8.5.2 Material recycling and energy 236
8.6 The Ultimate Problem for the Future of Mankind: The Flow of Free Energy 237
    8.6.1 Why energy? 237
    8.6.2 The prognosis for energy consumption 237
    8.6.3 How much energy is needed to produce the technological energy used by Man?. 239
    8.6.4 The future source of free energy. 240
    8.6.5 Not only the free energy sources are important but also the sinks! . 241
8.7 The Future Climate of This Planet. 242
    8.7.1 Will the terrestrial climate remain favourable?. 242
    8.7.2 The possibility of controlling the terrestrial climate 243
8.8 The Quality of Life. 243
8.9 What Conclusions Can Be Drawn Concerning the Future Development of Mankind? 244

CHAPTER 9

The Distant Future of Mankind – Terrestrial or Cosmic?

9.1 The Natural Constants and the Future of the Universe 247
    9.1.1 The very far future 247
    9.1.2 How stable are the natural laws and constants? 247
9.2 The Future Development of the Universe 249
9.3 The Future of the Galaxy and the Sun 250
    9.3.1 The stability of galaxies 250
    9.3.2 How stable is the cosmic neighbourhood of the Solar system? . . . 252
    9.3.3 How stable, how predictable is the Sun? 254
9.4 The Future of the Planet Earth 256
    9.4.1 The stability of the planet 256
    9.4.2 The fall of small cosmic objects and earthquakes 257
    9.4.3 The future terrestrial climate 258
9.5 The Possibilities for Mankind: Self-destruction, Self-isolation, Expansion 259
9.6 Human Colonies in Space – Possibility or Nonsense? 261
9.7 The Existence of Other Planetary Systems with Intelligent Life 262
    9.7.1 How many stars have planetary systems? 262
    9.7.2 How many planets having intelligent life could exist? 264
9.8 The Extraterrestrial Exchange of Information 265
9.9 Summary of the Limits of World Population Growth 267
9.10 Human Galactic Expansion and the Drake Limit 268
    9.10.1 The expansion velocity 268
    9.10.2 The energy need for cosmic journeys 270
9.11 Is It Really Impossible to Colonise the Galaxies? 272
9.12 The Very Distant Future; Mankind on This Planet 273
9.13 What Are the Conclusions Concerning the Distant Future of Mankind? 273

Bibliography 275
Index 281

CHAPTER 1

Matter and Energy. The Interplay of Elementary Particles and Elementary Forces

> Pluralites non es ponenda sine necessitatae. (Entities must not be needlessly multiplied.)
> W. Ockham
> (1280–1349)

> The essential mystery of the world is its comprehensibility.
>
> The human mind is not capable of grasping the Universe. We are like a little child entering a huge library. The walls are covered to the ceiling with books in many different tongues. The child knows that someone must have written these books. It does not know who or how. It does not understand the languages in which they are written. But the child notes a definite plan in the arrangement of the books – a mysterious order which it does not comprehend, but only dimly suspects.
> A. Einstein
> (1879–1955)

> Nature is not economical of structures – only of principles.
> A. Salam
> (1926–)

1.1 An Attempt to Describe the Natural World Using the Smallest Number of Elementary Phenomena

The aim of this book is to uncover the relationship between man and his Universe, to seek to discover his present position, and to draw some conclusions about his future.

At first glance these interactions between Man and Nature are so numerous, so different in character, so nebulous, and so little investigated that the likelihood of this book achieving its aim can be questioned from the first page.

1 Matter and Energy. The Interplay of Elementary Particles and Elementary Forces

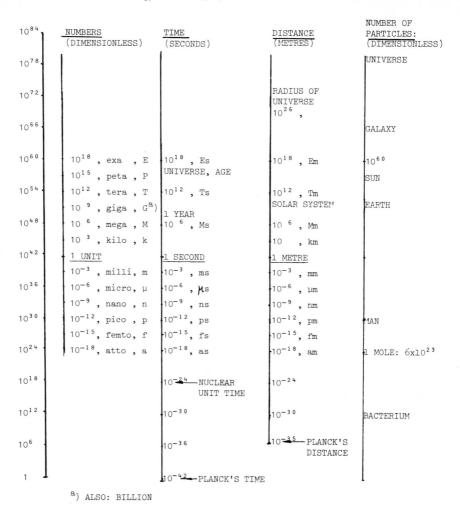

Figure 1.1. The scales of time, distance and particles.

It is to be hoped that at least part of the complex interplay of man and matter can be uncovered, if only by paying the price, at times, of oversimplification. It is to this end that the attempt to describe Nature, taking into account as few natural phenomena as possible, has been chosen. Already at this point, doubts and uncertainties arise. What is meant by "Nature"? How is "elementary" to be defined? At each stage of evolution considered here these terms will take on different meanings, but for the present we can use the phenomena shown below, which can adequately describe Nature as it is commonly understood:

– the particles, which are thought of as small "indestructible" spheres,
– the forces which govern the interaction of these particles,

1.1 Attempt to Describe the Natural World Using Minimal Phenomena

Table 1.1. Some Useful Conversion Factors

	Unit	Multiply by	To Obtain
Energy	1 eV (Electron volt)	1.602×10^{-19}	J (Joule)
	1 eV/Atom (Electron volt per atom)	96484.5	J/Mol (Joule per mole)
	1 J (Joule)	6.242×10^{18}	1 eV (Electron volt)
Mass	1 Mol	6.022×10^{23}	Atoms
	1 Electron mass	9.109×10^{-31}	kg
	1 Proton mass	1.672×10^{-27}	kg
Mass/Energy Relation	1 kg (Kilogram)	9.00×10^{16}	J
	1 Joule	1.111×10^{-17}	kg
	1 Proton mass	938.259	MeV (Mega-electron volt)
	1 Neutron mass	939.552	MeV (Mega-electron volt)
Temperature, Electromagnetic Waves	1 eV	11605	K (Kelvin); (Temperature)
	1 eV	1.241×10^{-6}	m; (Wavelength)
	1 eV	2.418×10^{14}	Hertz; (Frequency)
	1 Hz (Hertz)	4.135×10^{-15}	eV; (Energy)
	k Boltzmann constant	8.615×10^{-5}	$\frac{eV}{K}$ $\left[\frac{\text{Electron volt}}{\text{Kelvin}}\right]$
	k Boltzmann constant	1.380×10^{-23}	J/K

	Symbol	Unit	Value
Gravitational constant	G	$N \cdot m^2 \cdot kg^{-2}$	6.673×10^{-11}
Planck's constant	h	$J \cdot s$	6.62619×10^{-34}
Velocity of light	c	$m \cdot s^{-1}$	2.988×10^{8}
Elementary electric charge	e	C	1.602×10^{-19}
Mass of electron	m_e	kg	9.1095×10^{-31}
Fine structure constant	α	dimensionless	1/137.03595

- the natural constants (e.g., the velocity of light),
- the laws of Nature (e.g., mass–energy conservation).

From an aesthetic point of view and for theoretical cognitive reasons, the most satisfactory description of Nature would be one requiring only one kind of particle and one kind of force. Perhaps this description will one day be made.

Before proceeding with this discussion it may be useful to review the scale of numerical values which will be used, to give some idea of the basic properties of the natural world. This includes values for space, time, the strength of the elementary forces, and the number of particles in the Universe, and gives an opportunity to present the current language associated with these values (Figure 1.1 and Table 1.1).

1.2 General Foundations of the Physical Sciences

1.2.1 Some principles

In searching for and trying to understand the "laws of Nature", Man has evolved some properties which are concerned not with Nature itself, but with the way the search for those laws should be carried out and forming the basis for explaining them. These principles govern the research methods but not the objects of that research. For the purpose of this discussion the most useful are at least (Table 1.2):

- the principle of the Universality of the Laws (Newton's principle),
- the principle of Economy (Ockham's razor),
- the principle of Mediocrity (Copernicus' principle).

1.2.2 Some properties of the elementary phenomena are governed by very exact and strong laws of conservation

One of the most important achievements of physics was the discovery that the sum of mass and energy does not change. In each reaction the sum of energy and mass, before and after, is constant. Why is this conservation law neccessary? How is it derived from the other natural laws? Was it always so, everywhere?

Another example is the conservation of the sum of the electrical charges. In any process this sum also remains unchanged before and after. Why? What brings about this conservation law? Has it changed with time?

Some other examples are the laws of conservation of linear momentum and conservation of angular momentum. Other laws of conservation refer only to specific objects (and probably are not valid at very high temperatures):

Table 1.2. The General Principles

Principle of	Meaning of the Principle
Universality (Newton's)	All laws of nature, all phenomena which are known and valid here and now (on the Earth) are valid elsewhere in the Universe and in the past and in the future
Economy (Ockham's)	New entities (laws, phenomena, forces, particles) should not be postulated before the known "old" entities have been fully exhausted
Mediocrity (Copernicus's)	We are not in the centre of the worlds, we do not occupy a unique place (in space, in time). We are an average sample of the world – we are mediocre

1.3 Elementary Forces and Particles

Table 1.3. Some Prohibitions

Parameter	Prohibition, Meaning
Temperature, T	A temperature must be greater than absolute zero. $T > 0$ Kelvin (K)
Velocity, V	A velocity cannot be greater than 3×10^8 m/s. $V \leq C = 3 \times 10^8$ m/s
Heat, Spontaneous transfer, ΔH	Spontaneous transfer of heat from a lower temperature to a higher one without external work is not possible. If $\quad T'' > T'$, then $\quad T' \xrightarrow{\Delta H} T''$ is prohibited.

– *Conservation of "baryon number"*. The sum of the baryons remains constant. (The koino-baryon number $B = +1$ and the anti-baryon $B = -1$.)
– *Conservation of "lepton number"*. The sum of leptons remains constant. (The koino-lepton number $L = +1$ and the anti-lepton number $L = -1$.)

From other particles, such as mesons, being the middle-mass hadrons, and for photons, there is no such law of conservation. In any process the number of mesons or photons existing before and after need not be the same.

1.2.3 Prohibitions

In the Universe of natural laws there are some limits which go beyond the laws of conservation discussed above. These limits are best expressed as: process X cannot occur. The influence of these prohibitions cannot be overestimated. They play a vital role in natural phenomena (Table 1.3).

1.3 Elementary Forces and Particles

1.3.1 Elementary forces

A modern trend in physics seems to indicate a possibility of describing all known phenomena in the simplest way – in the form of one and only one force. We can call this elementary force the superunified force. The trouble is that this unique, still hypothetical force acts only in very hot conditions, where the temperature equals $\sim 10^{32}$ Kelvin, and over distances of only approx. 10^{-35} m. The question now arises, is it possible to achieve such an enormously high energy level and if so, for what length of time? The best answer is that it is not excluded that in the beginning of all things, in the beginning of all worlds, in the beginning of the Universe itself (including the "beginning" of space and time), during the very short time of approx. 10^{-43} second, the conditions were suitable for the action of the superunified force.

At some lower temperature, at longer distances and longer time periods, the superunified force decays into two other forces: the gravitational and the still hypothetical "grand unified force". In these conditions only very heavy, extremely short-lived, and very exotic particles can exist.

Moving to still lower temperatures, longer distances, and longer time periods, the grand unified force decays into two forces: the strong force, responsible for the nuclear phenomena, and the so-called "unified" force. The latter can be observed in the experiments of our biggest particle accelerators.

After a further decrease of temperature and an appropriate increase of distance and time, the unified force decays into two forces: the weak force, responsible for spontaneous changes in nucleons (protons and neutrons), and the electromagnetic force. The latter is the most extensively investigated force of Nature. It is responsible for the majority of phenomena observed in Nature, including chemical and biological processes (see Figure 1.2).

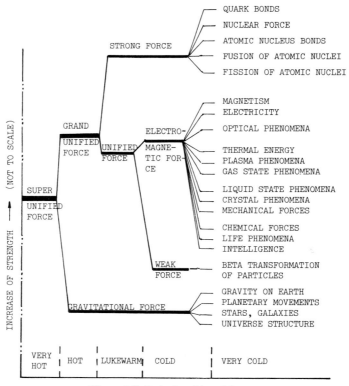

Figure 1.2. Relationship of forces.

1.3.2 Elementary particles

The description of Nature also includes the elementary particles. The properties and the number of different classes of elementary particles are much more complicated than those of the elementary force. Figure 1.3 shows in a somewhat simplified and to some extent arbitrary way the relationship between the elementary particles.

From this unified description of the elementary forces and elementary particles we can conclude that all four of the elementary forces which we know from practical experience and from our experiments are mutually connected, and inherently coupled (Figure 1.4).

The unity of Nature, and the unity of the whole Universe, is the most important conclusion from this consideration. It is possible that the known forces and laws govern the whole Universe, everywhere, and at any time, including the very early epoch of the beginning of the Universe. The material and energetic bases for life are also common in the Universe. All the intelligence in the whole Universe may even be based on a common foundation.

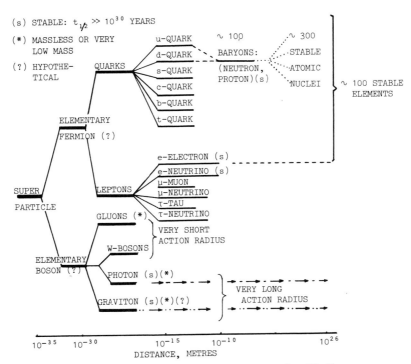

Figure 1.3. Relationship of particles (simplified).

8   1 Matter and Energy. The Interplay of Elementary Particles and Elementary Forces

		ELEMENTARY FORCES			
	FORCE DOES NOT ACT	SHORT ACTING RADIUS $< 10^{-1.5}$ m		LONG ACTING RADIUS $> 10^{25}$ m	
		STRONG	WEAK	ELECTRO-MAGNETIC	GRAVI-TATION
	STRENGTH OF FORCE	~ 1	$\sim 10^{-12}$	$\sim 10^{-2}$	$\sim 10^{-40}$
	ATTRACTION REPULSION	ATTRACTION –	?? --	ATTRACTION REPULSION	ATTRACTION –
PARTICLES / LIGHT AND HEAVY	QUARKS ↓↓↓ HADRONS	QUARKS ↓↓↓ HADRONS	QUARKS ↓↓↓ HADRONS	QUARKS ↓↓↓ HADRONS	QUARKS ↓↓↓ HADRONS
	LEPTONS		LEPTONS	LEPTONS CHARGED	LEPTONS
"MASSLESS" PARTICLES	PHOTON			PHOTON	PHOTON
	GRAVITON				GRAVITON

Figure 1.4. All four elementary forces.

1.4 Elementary Particles

1.4.1 "Bricks" and "mortar"

The many different types of hadrons are far from being accepted as distinct elementary particles. This unsatisfactory state can be improved by considering the so-called quark model. This model permits a surprisingly good explanation for the existence of so many forms of hadrons – approx. 100 koino-hadrons and 100 anti-hadrons. Today's version of this hypothesis assumes the existence of different quarks and anti-quarks of the so-called "first", "second", and "third" generation (Table 1.4).

The similarity with the six koino-leptons and six anti-leptons is very evident. However, there are essential differences among them.

Some leptons are stable and can be observed and measured in each of their six forms. Up to now the quarks remain hypothetical. There are even theories which suggest that quarks by their nature must remain unobservable particles, except for combinations of quarks such as a quark and anti-quark together forming a meson, and combinations of three koino-quarks together making a koino-baryon (e.g., proton) or three anti-quarks forming an anti-baryon (e.g., antiproton).

1.4 Elementary Particles

Table 1.4. Elementary Mediators and Particles

		"Mortar" That Binds the Bricks: Mediators				"Bricks" of Which the World Is Built: Particles	
		Bosons				Fermions	
Properties of the Objects		Many Particles in a Single State				Only One Particle in a Single State	
Example		Photon γ	Gluons g	Graviton	W-Bosons W^+, W^-, Z^0	Leptons	Quarks
Electrical charge		0	0	0	$+1; 0; -1$	$+1; 0; -1$	$+\frac{1}{3}; -\frac{1}{3}; +\frac{2}{3}; -\frac{2}{3}$
Mass		0	0	0	85 GeV	> 0	Heavy
Kind of Matter	"Ordinary" (first generation)	Mediator of the Electromagnetic Force	Mediator of the Strong Force	Mediator of the Gravitational Force	Mediator of the Unified Force	Electron (e)	Up quark (u)
						Electron-neutrino (v_e)	Down quark (d)
	"Strange" (second generation)					Muon (μ)	Strange quark (s)
						Muon-neutrino (v_μ)	Charmed quark (c)
	"High energy" (third generation)					Tauon (τ)	Top quark (t)
						Tau-neutrino (v_τ)	Bottom quark (b)

1.4.2 Creation of the elementary particles

The laws of conservation do not exclude the possibility of creation of elementary particles or their eventual extinction. These laws only regulate those creation and destruction processes. The best-known example of the birth of an elementary particle is the emission of a photon from an excited or "hot" atom. The process can be described as:

$$A^* \rightarrow A + \gamma \quad (A^*: \text{excited atom}; \gamma: \text{photon}).$$

Of course, here the conservation laws – mass–energy, momentum, and electrical charge – are still valid. There is, however, no conservation law for photons, therefore a photon is created. The photon can be destroyed in two different ways:

- firstly, by absorption in an atom:
  $\gamma + A \to A^*$ (the reverse of the earlier process) which gives an excited atom, A^*;
- secondly, when the photon has a high enough energy – greater than 1.02 MeV – the following process is possible:
  $\gamma \to e^- + e^+$ (e^+ = positron = anti-electron);
                 (e^- = electron = koino-electron);
                 (γ = photon).

(See Figure 1.5.) This process is known as "pair production".

Here again the reverse process is well-known and is called "pair annihilation":
$$e^+ + e^- \to \gamma + \gamma.$$

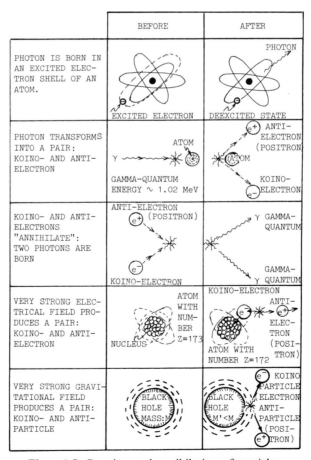

Figure 1.5. Creation and annihilation of particles.

1.4 Elementary Particles

The creation of massive particles is not limited to the leptons. The creation of hadron pairs has also been observed:

$$\gamma \to p^+ + p^- \qquad (p^+ = \text{koino-proton};$$
$$p^- = \text{anti-proton}).$$

It is not neccessary to repeat all the conservation laws that are valid in such cases. In this pair production, not only a koino-particle, but also an anti-particle, is produced. Is the reverse process possible?

Finally, the rather hypothetical processes which recently are receiving intense investigation can be mentioned:

$$\text{Gravitational force} \to \text{koino-particle} + \text{anti-particle}.$$

This process is possible only in a very strong gravitational field. Such a field is supposed to exist in two cases: in the neighbourhood of a "black hole" and in the very early stages of the evolution of the Universe (see Chapter 2).

1.4.3 "Life" and "death" of elementary particles

Some of the elementary particles, including the quark-agglomerates, the hadrons, are very stable, while others are unstable.

What is meant by a "stable" particle? Does it mean that a stable particle is stable forever, under all conditions and circumstances? Under all pressures, temperatures, or other influences? It is necessary to look at the concept of stability, which is here of the highest importance, in rather more detail.

The first question is: what kind of elementary particle is stable in free space without being influenced by other phenomena, such as temperature, pressure, forces, etc?

The answer is that under these conditions only a very few elementary particles are stable, that is, remain unchanged for a long time. These particles are the massive ones, the proton and electron, and the massless photon and graviton, the latter two always having a velocity of 3×10^8 m/s in vacuum.

In a particular experiment, involving on the order of 10^{33} protons and electrons with a duration of 10^7 seconds, no change of particles was observed. Hence it may be deduced that, even after 10^{32} years for protons and 10^{21} years for electrons, more than half of these particles will remain unchanged.

What a great stability! The Universe itself is supposed to be something less than 2×10^{10} years old. If the measurement of the stability of protons is correct, giving a half-life greater than 10^{30} years, then within the lifetime of the Universe only about one in 10^{19} of all protons would have decayed. There is no doubt that these are really stable particles. The modern grand unified theory predicts a half-life for protons of $\gtrsim 10^{32}$ years.

The enormous stability of the proton, which is the lightest of all baryons, has been promoted to the status of a law: the conservation of baryon number.

This is the expression of the very small probability of the spontaneous decay of a proton, according to the grand unified theory:

$$\text{proton}^{(+)} \xrightarrow[(t_{1/2} \sim 10^{32} \text{ years})]{\text{grand unified force}} \text{meson}^{(0)} + \text{positron}^{(+)}.$$

The reason why the law of conservation of baryon number should be so exactly followed can be explained by the very high energy needed for this proton decay, on the order of 10^{15} GeV. Suffice it to say that the existence of the Universe as one observes it is the result of this law.

1.5 The Existence of Atomic Nuclei Is Due to the Forces of Attraction Between Their Nucleons

1.5.1 The weak force limits the number of stable hadrons

Of the approximately 100 hadrons, only two are stable or quasi-stable. These are the proton and the neutron. In the free state the particles have the following properties:

	Nucleons	
	Proton	Neutron
Baryon number	+1	+1
Spin	1/2	1/2
Mass (MeV/c^2)	938.259	939.552
Electric charge	+1	0
Half-life (seconds)	$\leqslant 10^{39 \pm 1}$	~ 636

The proton and neutron, being constituents of the nucleus, are sometimes referred to by the general name nucleons.

All hadrons, as well as the leptons, are ruled by the weak force. The weak force is a universal force which always acts except in forbidden regimes. The boundary of its action is set by some laws of conservation; in the case of the neutron-proton the boundary is a combination of the following:

mass-energy conservation,
conservation of electric charge,
conservation of baryon number,
conservation of lepton number.

Thus the only process for transforming a neutron into a proton which obeys all four laws of conservation is:

$$\text{neutron}^{(0)} \xrightarrow[\text{beta-decay}]{} \text{proton}^{(+)} + \text{electron}^{(-)} + \text{antineutrino}^{(0)} + \text{free energy}.$$

1.5.2 Strong force binds the nucleons together

The nucleons – that is, the neutron and proton – are influenced by the strong force (also known as the nuclear force). It manifests itself only as a force of attraction. If this were the only force acting on the nucleons the most likely collection of particles would be the polyneutrons, a sphere containing a large number of neutrons. Such a collection of neutrons is not impossible, except that the number must be around 10^{57} neutrons, weighing some 10^{30} kg. Here the gravitational force also plays a role, affecting the stability of the conglomerate. This is the neutron star (Chapter 3).

Where the number of neutrons in the mass is significantly smaller, the weak force, in addition to the strong force, can also play a part. Since the mass of the bonded protons approximately equals the mass of the bonded neutrons, some of the neutrons are able to transform into protons due to the weak force (via beta-minus decay).

The resulting conglomerate now includes not only neutrons but also protons. A rather significant new phenomenon now comes into play. The protons carry a positive electric charge, and electrically charged particles fall under the influence of the electromagnetic forces. This elementary force is manifested not only by attraction but also by repulsion. Too large a number of positively charged protons would therefore result in the fragmentation of this conglomerate, due to their mutual repulsion. This process is known as the fission of the atomic nucleus.

The consequence of this interplay of the strong force with the electromagnetic repulsion force is that certain combinations of protons and neutrons are more stable than others. These special "totals" of groups of protons and neutrons, called magic numbers, are:

$$2, 8, 20, 28, 50, 82, 114, 126.$$

Stable spatial configurations of particles also occur on the atomic level. This is the realm of the chemical elements.

The spatial structure of the negatively charged electrons, orbiting around the positively charged atomic nucleus, is particularly stable when the electrons have the following numbers:

$$2, 10, 18, 36, 54, 86, 118.$$

These are the atomic numbers (Z) of the relatively stable (nonreacting) chemical elements, known as the noble gases:

$Z =$ 2, helium
$Z =$ 10, neon
$Z =$ 18, argon
$Z =$ 36, krypton
$Z =$ 54, xenon

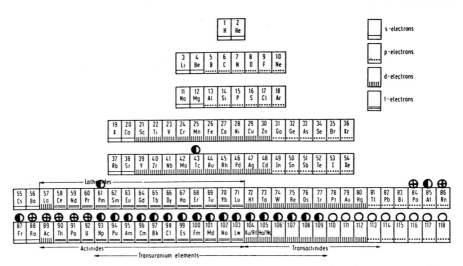

Fig. 1.6. Periodic table of elements. ⊕ : Radioactive (unstable) elements, occuring naturally in the Earth's crust; ◐ : radioactive (unstable) elements, man-made; ○ : still not discovered.

$Z = $ 86, radon
$Z = 118$, not yet detected; a noble-gas element.

Figure 1.6 shows the periodic table of the elements.

1.5.3 Binding energy of a nucleon

This section shows what happens when we try to create from a simple neutron or proton a cluster of such particles by "glueing" on more neutrons or protons (nucleons).

Figure 1.7 shows the results of adding nucleons. In some cases it is possible to arrive at a rather stable conglomerate, a stable atomic nucleus. More frequently, however, the collection of nucleons which has been stuck together is unstable and changes spontaneously into another configuration, without altering the total number of nucleons. This is possible when a neutron transforms into a proton or vice versa. This is the weak force transformation, also called the beta-transformation.

In still more cases the conglomerate is very unstable and breaks apart or fissions, forming smaller particle clusters, or fission products. In the most unstable cases, the nucleons cannot stay together longer than 10^{-23} s, which is the shortest period observable.

What is the reason for these effects? They are all due to the action of the weak force.

1.5 The Existence of Atomic Nuclei Is Due to the Forces of Attraction

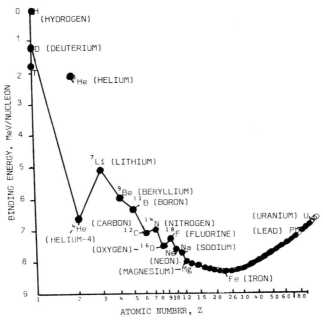

Figure 1.7. Binding energy per nucleon for stable nuclides.

The weak force determines the number of stable nuclides. If one tests all combinations of protons and neutrons, the following picture emerges. Combining protons up to the magic number $Z = 82$ and the neutrons from zero up to $N = 126$:

- The number of all "possible" nuclides is $82 \times 126 = 10332$.
- The number of stable nuclides – that is, not transformed by the weak force – is only 272, that is, only 2.7 percent of all "possible".
- The number of nonstable nuclides transforming by beta-transformation is estimated to be approx. 1500, that is, six times more than the stable nuclides.
- The rest, the very unstable nuclides, can be said "not to exist".

1.6 Matter and Free Energy – The Intimate Connection

The unit of energy in the International System (SI) is the joule (J), although on the microscopic scale the electron volt (eV) may be used. Whereas the unit of mass is the kilogram (kg), the close connection between mass and energy

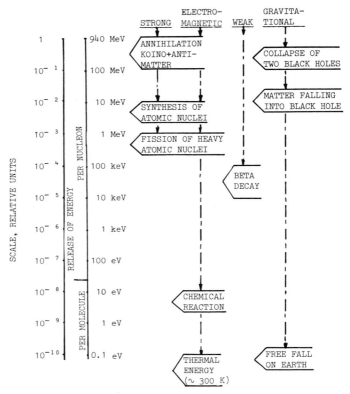

Figure 1.8. Transformation of mass into free energy due to elementary forces.

requires a common unit, one which has already been used in the earlier parts of this chapter:

$$\text{energy} = \text{mass} \times c^2$$
$$(\text{Joule}) = (\text{kg}) \times (3 \times 10^8 \text{ m/s})^2.$$

(Figure 1.8 gives the same relationship for the microscopic units.)

From this it can be seen that in some processes it is possible to transform some of the massive particles into energy, which then appears as either the kinetic energy of the particles or as another form of energy, for example, photons.

The amount of energy so produced from various processes is shown in Figure 1.8. Thus it can be deduced that:

– The reaction converting most material into free energy is the annihilation process, that is, the collision of a koino-particle with its anti-particle. Almost the whole mass can be transformed into free energy mostly as photons (gamma quanta) and as kinetic energy of koino- and anti-neutrinos.

- A rather effective way of transforming matter into free energy is the collapse of two black holes. Here about half of the mass can be transformed into free energy in the form of gravitational waves (gravitons).
- Capture of massive particles by the black hole is less effective, and only part of the mass can be transformed.

Both these last processes are the manifestation of the gravitational force.

- The process of nuclear fusion of light nuclides, mostly protons and deuterons, resulting in the syntheses of helium, converts about 1 % of the mass into energy.
- Somewhat less effective still is the fission process of the heavy nuclides, e.g., uranium, which releases about 0.1 % of the mass as energy.
- The beta transformation, an effect of the weak force, transforms 1/10,000 of the mass into energy.
- Finally, the chemical reaction, which is governed by the electromagnetic force, transforms only 10^{-8} of the mass into free energy.

1.7 What Are the Conclusions for the Future Development of Mankind?

From all these questions discussed in this chapter:

- the "eternal" and universal laws of Nature and the importance of the natural constants,
- the interplay of the elementary forces and the elementary particles,
- the birth, life, and death of the particles, the elements,
- the coupling of energy and matter and the resulting processes of ordering the material objects,

the following conclusions can be drawn for the decisive problem of this book, the future development of mankind:

- The scientific description of Nature evolves toward a more complete picture in which different parts result in one harmonic entity.
- The increasingly integrated scientific model of the world allows the preparation of a prognosis of the future development of Nature on a cosmic and planetary scale.

It is trivial to add that these are only the physical bases for the further evolution of mankind. The ultimate choice of the future way is doubtless one to be made by Man himself.

CHAPTER 2

The Universe: How Is It Observed Here and Now? Its Past and Possible Future

God is complicated but wicked he is not ... he doesn't play dice with us.
Der Herrgott ist kompliziert aber boshaft ist er nicht ... und spielt mit uns nicht mit Würfeln.

A. Einstein
(1879–1955)

The Universe is everything: both living and inanimate things, both atoms and galaxies, and if the spiritual exists as well as the material, of spiritual things also; and if there is a Heaven and a Hell, Heaven and Hell too; for by its very nature the Universe is the totality of all things.

F. Hoyle
(1910–   )

Modern scientific man has largely lost his sense of awe of the Universe. He is confident that, given sufficient intelligence, perseverance, time, and money, he can understand all there is beyond the stars. He believes that he sees here on earth and in its vicinity a fair exhibition of nature's laws and objects, and that nothing new looms "up there" that cannot be explained, predicted, or extrapolated from knowledge gained "down here". He believes he is now surveying a fair sample of the Universe, if not in proportion to its size – which may be infinite – yet in proportion to its large-scale features. Little progress could be made in cosmology without this presumptuous attitude. And nature herself seems to encourage it, as we shall see, with certain numerical coincidences that could hardly be accidental.

W. Rindler
(1924–   )

2.1 What Is the Universe?

2.1.1 A definition of the Universe

We will attempt here to describe the Universe using the well-established language of physics.

To make this description very little is needed – four elementary forces: strong, electromagnetic, weak, and gravitational; and four classes of elementary particles: hadrons (probably made up of six kinds of quarks), six types of leptons, the photon, and the graviton (still a hypothetical particle). With these building blocks the Universe will be described. No other "ad hoc" forces or particles are required. The aim will be to follow the principle of "Ockham's Razor":

"Things should be kept as simple as possible"

(Figure 2.1).

2.1.2 Beginning of the Universe

Almost all generally accepted models of the Universe postulate a "beginning" of the Universe and further evolution, including rather unusual properties of the young Universe. All these models give a more or less exact and detailed description of the very beginning, the so-called "Big Bang", and the following expansion (Figures 2.2 and 2.3).

In the beginning the Universe was very hot, very dense, and very small. Here are some details of the Universe's evolution.

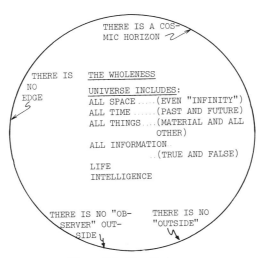

Figure 2.1. The Universe.

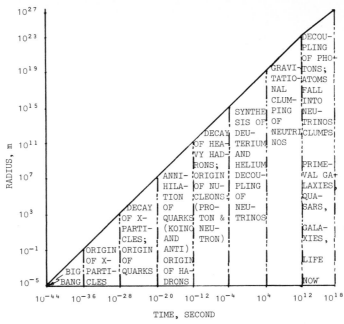

Figure 2.2. Expansion of the Universe.

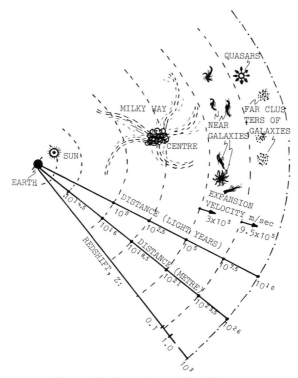

Figure 2.3. Expansion of the Universe.

2.2 Expansion of the Universe

2.2.1 The red shift

The expansion of the Universe is demonstrated by the observation that the known characteristic wavelengths of photon emissions from specific elements, such as sodium and hydrogen, when arriving from distant objects are shifted towards a lower photon energy. For visible light this means a shift toward the red end of the spectrum. This is called the "red shift".

All explanations of this phenomenon, with the exception of the Doppler effect, have been shown to be invalid because of inconsistencies with other observable facts. Only the Doppler effect, due to the movement of the emitting objects away from the observer, fits the known facts.

2.2.2 The five eras of the Universe

The best-established model of the evolution of the Universe includes the following eras (see Table 2.1):

- very hot, directly after the Big Bang, also called the Planckian era,
- hot,
- lukewarm,
- cold,
- very cold, the era in which we are living.

2.3 What Is Known About the Universe Today?

2.3.1 The average composition of the Universe

To begin with the make-up of a mean sample of the contents of the Universe will be helpful. Assume a homogeneous mixture of all constituents from which a sample volume of one cubic metre is taken. If the contents could be fully analysed it would be found to contain (Figure 2.4):

a) Particles:
- hadrons in the form of stable nucleons: protons and semi-stable neutrons. Total amount: approx. 0.06 nucleons per m^3 (approx. 1 per $16 m^3$);
- leptons in the form of electrically charged stable electrons: the number is probably exactly equal to the number of protons, i.e., approx. 0.06 per m^3;
- leptons in the form of very light neutrinos, electrically neutral. The number is not well known, but it is about 500 million per m^3. (No differentiation between koino- and anti-neutrinos made here.)

Table 2.1. Five Eras of the Universe's History

Time (seconds)	10^{-44}	10^{-36}	10^{-28}	10^{-20}	10^{-12}	10^{-4}	10^{+4}	10^{12}	10^{18}
Temperature (Kelvin)	10^{32}	10^{28}	10^{24}	10^{20}	10^{16}	10^{12}	10^{8}	10^{4}	10^{0}
Radius (metres)	10^{-5}	10^{-1}	10^{3}	10^{7}	10^{11}	10^{15}	10^{19}	10^{23}	10^{27}
Era of the Evolution of Universe		Very hot	Hot	Lukewarm	Cold				Very cold
Epoch						Quarks	Hadrons	Leptons	Photons	Stars
Forces		Super unified								
			Grand unified	Strong						
				Unified	Weak					
					Electromagnetic					
					Gravitation					
Events		↑ Big Bang								
			↑ Decoupling of gravitation							
				↑ Decoupling of strong force						
					↑ Decoupling of weak force					
				Decoupling of neutrinos ↑						
						Decoupling of photons ↑				
						End of "Fireball" ↑				
							Life ↑			
								Intelligence ↑		

b) Mediators:
- photons, which are very numerous at 500×10^6 m³ and mainly at an energy of 0.2 meV (milli-electron volts) corresponding to blackbody radiation at a temperature of 2.9 Kelvin. Only a few photons are at the level of infrared, visible, or ultraviolet radiation;
- gravitons: these are suspected but so far remain unobservable ($> 10^8/m^3$??).

c) Forces between particles and mediators:
- the most important, the gravitational force, influencing all particles without exception;
- the next is the electromagnetic field, being a manifestation of the electromagnetic force;
- the other two forces, the strong (nuclear) force and the weak force, can be detected only in the neighbourhood of the particles themselves.

2.3 What Is Known about the Universe Today?

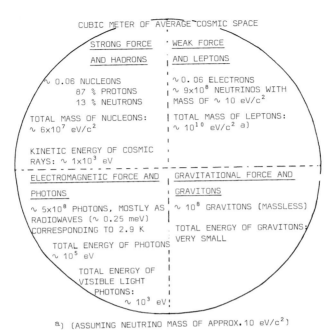

Figure 2.4. Contents of one average cubic metre of cosmic space.

In total it can be claimed that the mean sample of the Universe contains all the elementary forces so far known: gravitational, electromagnetic, weak, and strong; and both elementary classes of particles – leptons and hadrons. The claim is not trivial and means further that there are no other particles or forces involved in the terrestrial sphere which cannot be reduced to those named.

It is not necessary when describing the Universe to use "ad hoc" descriptions of forces or particles.

To return to the estimates of the average composition of the present Universe, the results indicate the following points:

a) The Universe is rather empty, about 1 atom per $16\,m^3$. This is $\sim 10^{27}$ times less than the number of molecules in a cubic metre of terrestrial air. The average density of cosmic matter is very small and equal to only $2 \times 10^{-26}\,kg/m^3$. However, it must be said that this value is still somewhat uncertain and may be a factor of 10 lower.

b) The Universe is not only empty but transparent to the transmission of light. It is possible for a photon of light to traverse the entire Universe without being absorbed or deflected. This allows direct observation of the distant parts of the Universe.

c) The Universe is cold. The majority of photons have an energy of only $0.2\,meV$ and a wavelength of $\sim 1\,mm$. Such radiation corresponds to an emmission of a black body with a temperature of only 2.9 K. For example, the Sun, which has a surface temperature of 5785 K, emits photons with an energy of $\sim 2\,\text{eV}$, corresponding to a wavelength of $\sim 600\,\text{nm}$.

d) The mass of the Universe is probably due mainly to neutrinos, probably over 9/10 of the total mass of the Universe (assuming neutrinos are massive).

e) The total mass of the present Universe resulting from the mass of neutrinos and partially of baryons equals $2.5 \times 10^{54}\,\text{kg}$.

All remaining components – electrons, photons and gravitons – do not influence this result.

2.3.2 Chemical composition of cosmic matter

The next question is: what chemical elements are to be found in the material which exists in atomic form at the present time?

Many observational methods exist which permit a chemical analysis of cosmic objects. The results are rather amazing (Table 2.2):

Table 2.2. Cosmic Abundance of Elements

Order	Element	Symbol	Atomic Number Z	Relative Abundance
1	Hydrogen	H	1	920,461
2	Helium	He	2	78,344
3	Oxygen	O	8	608
4	Carbon	C	6	305
5	Nitrogen	N	7	84
6	Neon	Ne	10	77
7	Iron	Fe	26	37
8	Silicon	Si	14	30.5
9	Magnesium	Mg	12	24.3
10	Sulphur	S	16	15.6
11	Argon	Ar	18	5.8
12	Aluminium	Al	13	2.2
13	Calcium	Ca	20	2.2
14	Nickel	Ni	28	1.8
15	Chromium	Cr	24	0.6
16	Phosphorus	P	15	0.3
17	Chlorine	Cl	17	0.36
18	Manganese	Mn	25	0.20
19	Potassium	K	19	0.08
20	Fluorine	F	9	0.04
21–81	All others up to Thorium	–	3, 4, 5 and 29 up to 89	0.04
82	Thorium	Th	90	0.0000002
83	Uranium	U	92	0.0000002
			1–92	1,000,000

2.3 What Is Known about the Universe Today?

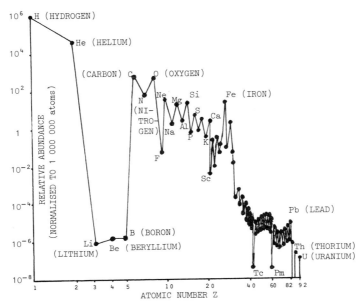

Figure 2.5. Cosmic abundance of elements.

- Cosmic matter contains probably all stable and some quasi-stable elements. Only very short-lived elements cannot be measured in very distant objects. Also some unstable radioactive elements which do not exist on Earth in the natural state, such as technetium (atomic number $Z = 43$) and promethium ($Z = 61$), have been detected in some stars.
- For each 1000 atoms of cosmic matter, approx. 920 are hydrogen atoms, the lightest and simplest of all elements with atomic number $Z = 1$.
- For each 1000 atoms of cosmic matter, approx. 78 are atoms of helium, the second most abundant element.
- All other ~90 stable and semi-stable elements make up the remaining 1 per 1000 atoms.
- Of these heavier elements, most abundant are oxygen, carbon, nitrogen, and neon (Figure 2.5).

2.3.3 Composition of photons

The number of photons in the Universe is as large as the number of neutrinos. For every baryon there are probably ~ 10^9 photons. Most, some 995 in every thousand, have an energy corresponding to a wavelength of ~ 1 mm. This is the region of radio waves; the mean energy is about 0.2 meV. For comparison, the number of photons in the visible light region emitted from stars and other objects amounts to about one thousand per cubic metre, the mean energy per photon being 2 eV with a wavelength of ~ 600 nm.

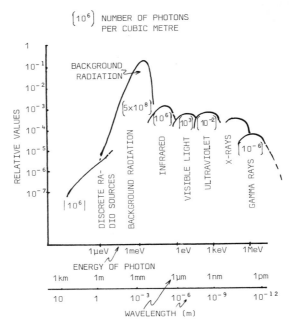

Figure 2.6. Spectrum of radiation in cosmic space.

A very important feature of this ancient radiation (formed long ago) is that it arrives on Earth from all directions of space with astonishing isotropy. Exact measurement shows a deviation of less than 0.01 percent from the mean in any direction. The second unusual feature of this radio-wave radiation is the shape of the spectrum. This corresponds exactly to radiation from a blackbody having a temperature of 2.9 K (Figure 2.6).

2.4 The Universe as a Whole

Without giving the observable facts which form the basis for the physical description of the Universe, it can be characterised as a whole in the following ways (Figure 2.7).

a) The Universe is big. The limit of observation, the cosmic horizon, is about 2×10^{26} m distant. Light moving with a velocity of 3×10^8 m/s reaches the observer from the horizon after $(2 \times 10^{26})/(3 \times 10^8 \text{ m/s})$ or $\sim 6.7 \times 10^{17}$ s, or ~ 21 gigayears. The volume of the Universe up to the cosmic horizon is about $\sim 10^{80}$ m^3.

b) As will be discussed later, the cosmic horizon is limited by the state of matter in the Universe when it was one million years old. With optical observation

2.4 The Universe as a Whole

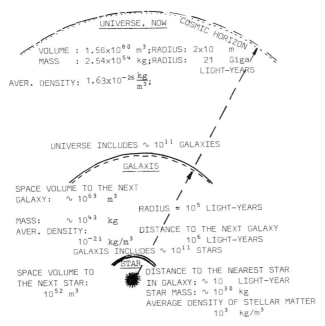

Figure 2.7. The structure of the Universe at the present time.

it is not possible to "see" further back, since only after 1 million years did the Universe become transparent.

c) The Universe includes about a billion clusters of galaxies. The average cluster includes about one thousand galaxies (the existence of galaxies in clusters is still under dispute). The galaxies are the "atoms" of the Universe.

d) The galaxies are moving away from each other. This expansion of the intergalactic space is a process which has been occurring for a very long time. The velocity of expansion increases with the distance from the observer at a rate of about 14 km/s for each million light years of distance. This value is deduced from numerous observations but is partially based on a more or less arbitrary assumption of "Hubble's constant", which, at the present time, equals:

$$H_0 \simeq 14 \frac{\text{km/s}}{\text{mega-light-year}} = 1.5 \times 10^{-18} \text{ s}^{-1}.$$

e) The Universe as a whole seems to be:
- homogeneous, that is, having the same properties, the same density, the same type of objects everywhere,
- isotropic, that is, having the same properties (e.g., expansion and velocity) in all directions.

2.5 The Future of the Universe

The uncovering of the history of the Universe can be considered as one of the triumphs of human reasoning.

Who, however, can restrain himself from being a prophet and trying to extrapolate to the future development of the Universe based on the present state of knowledge? Possible developments are:

a) The Universe continues to expand forever.
b) The Universe expands but reaches a limit after which it begins to contract (Figure 2.8).

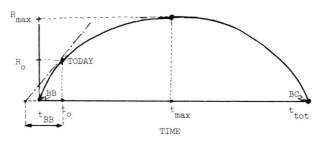

Figure 2.8. Model of a closed universe. t_H = Hubble's time = 21 Gyr; BB = Big Bang; BC = Big Crunch.

Table 2.3. Universe in the Far Future According to the Model of a Closed Universe (c = velocity of light: 3×10^8 m/sec; G = gravitational constant: 6.67×10^{-11} N · m²/kg²)

Parameter	Symbol	Calculation	Unit	Value
Universe at Maximum Expansion of Radius				
Radius	R_{max}	$R_{max} = 2R_0$	m	3.98×10^{26}
Age	t_{max}	$t_{max} = \pi \cdot t_H$	sec	2.09×10^{18}
			Gigayear	66
Universe at Big Crunch				
Age	t_{tot}	$t_{tot} = 2 \cdot t_{max}$	Gigayear	132

2.6 What Conclusions Can Be Drawn for the Future Development of Mankind?

Some conclusions may be drawn about the central theme of this book – the future development of mankind. These conclusions are based on the following factors:

- the present state of the Universe, the average properties of cosmic matter and energy;
- the structure of the Universe including the galaxies and the continuing expansion;
- the model of the Universe's expansion;
- the coupling between some average properties, such as density of the Universe and trends of future development.

The conclusions are as follows:

- Although the present model of the evolving Universe contains many doubtful and uncertain aspects, enough is certain for some predictions to be made with confidence.
- The Universe is likely to go on evolving at least in the coming tens of billions of years in the same direction as the last ten billion years.
- The influence of the Universe on mankind is unlikely to pass through any dramatic phase.
- An assured developmental path for mankind is likely to exist in the future and extend even into the far distant future.

Only the external physical influences are considered here – a far bigger influence on mankind's development comes from Man himself!

CHAPTER 3

The Origin and Nuclear Evolution of Matter

> It seems probable to me that God in the Beginning form'd matter in solid, massy, hard, impenetrable, moveable particles, of such Sizes and Figures and with such other Properties and in such proportion to Space, as most conduced to the End for which he formed them.
>
> Isaac Newton
> (1642–1727)

3.1 The Creation of the Elementary Particles in the Very Early Universe

3.1.1 Unknown phase: Era of superunified force (Planckian Era or Very Hot Era)

We begin the story of the Universe when it is only 10^{-44} second old. The temperature is approx. 10^{32} K and the density is enormously high, $\sim 10^{96}$ kg/m^3. The radius of the Universe is extremely small.

It has been speculated that, in these conditions, only one primordial force is present, but up to now there has been no established description of it. This hypothetical force is called the "superunified force" (also "supergravity") and probably acts on all possible particles although the particles which exist in these conditions are unknown. More precisely, there exists no idea, no hypothesis concerning these primordial particles ("superparticles"?) and the elementary "superunified" force.

This era is called the Planckian Era or Very Hot Era (see Table 2.1).

3.1.2 Era of grand unified force (Hot Era)

After approximately 10^{-36} second the temperature decreases to approx. 10^{28} K. This corresponds to an energy of approx. 10^{15} GeV (remember: 1 eV is equivalent to 11,605 K; see Table 1.1). The Universe expands and the radius becomes larger; the density decreases to $\sim 10^{80}$ kg/m^3.

For these conditions there exists a theoretical basis: the so-called "grand unified theory" (GUT). According to this theory, there exist two different forces:

– gravitational force,
– grand unified force.

The gravitational force in this era is much weaker than the grand unified force (see Figures 1.2 and 1.4).

The very heavy particles emerge, such as X-particles. There are two kinds of these very heavy X-particles (with mass of $\sim 10^{-15}$ GeV):

– koino-particles: X,
– anti-particles: $\tilde{X}$.

These particles are very unstable and in a very short time they decay. The most surprising hypothesis is that these X-particles decay in a rather asymmetric manner:

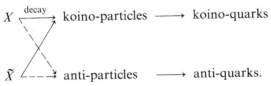

Here

$\longrightarrow$ for major decay mode,
$-\!-\!\!\rightarrow$ for minor decay mode.

As a result, more koino-particles than anti-particles are formed. The matter in the Universe is no longer symmetrical.

3.1.3 Era of unified force (Lukewarm Era)

The Universe expands further: the radius is now of an order of magnitude of metres. The temperature decreases significantly and achieves the level of 10^{20} K. The Universe is 10^{-20} second old. The "Lukewarm Era" begins; the grand unified force splits into two forces (see Figure 1.2):

– strong force,
– unified force (called also "electroweak force").

From the foregoing "Hot Era" there exist the gravitational force and the quarks (koino-quarks and anti-quarks). The annihilation of quarks results in photons:

$$\text{koino-quark} + \text{anti-quark} \xrightarrow{\text{annihilation}} \text{photons.}$$

Only one koino-quark in 10^9 quarks (anti + koino) survives this annihilation process.

3.1.4 Cold Era and Very Cold Era

The Universe is now 10^{-10} second old. The temperature decreases to 10^{15} K, which corresponds to energy of approx. 100 GeV.

The unified force ("electroweak" force) splits into the two following forces:

- electromagnetic force,
- weak force.

From earlier epochs there exist:

- gravitational force,
- strong force.

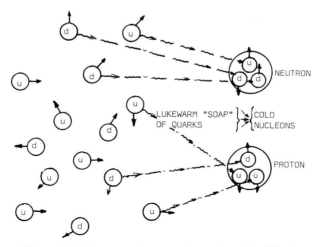

Figure 3.1. From quarks to nucleons (very simplified).

3.2 Evolution of the Elementary Particles

Now the four elementary forces exist which we know from our everyday experience.

The koino-quarks (and a small proportion of the anti-quarks?) which survive the annihilation begin to clump together – the baryons are in the making.

Figure 3.1 shows schematically the synthesis of the most stable baryons in the "quark soup": the nucleons.

3.2 Evolution of the Elementary Particles.
A Very Rapid Development in the First Seconds of the Universe

3.2.1 Beginning of the Cold Era: Evolution in the "Hadron Epoch"

It is perhaps better to leave the uncertain ground of the hypotheses dealing with the creation of matter and move to the more soundly based theories of the evolution of matter in the "Hadron Epoch".

We begin arbitrarily with the very young Universe, having an age of ~ 100 microseconds (one ten thousandth part of a second). The Universe at this point is rather small, having a radius of approx. 10^{15} m, that is, one hundred times larger than the distance of the Sun to Pluto, the outermost planet of our Solar System. The density is approx. 10^{19} kg/m^3 and the temperature 10^{12} K. At this temperature the mean energy of the photons is approx. 100 mega-electron volts (100 MeV).

What is the interpretation of these basic parameters? The following is a simplified answer:
a) The density of 10^{19} kg/m^3 corresponds to the density of the nucleons themselves (protons and neutrons). This means that at this moment the nucleons, or rather the hadrons, are so closely packed that no free space exists between them. This corresponds to a state in which each baryon (e.g., proton or neutron) occupies a set space and no other baryon can occupy it at the same time. This rule is not valid for the mesons (see Chapter 1).
b) The temperature of $\sim 1 \times 10^{12}$ K corresponds to the mean energy of photons of 100 MeV, but due to the quasi-equilibrium state some of the photons reach energies in excess of 1000 MeV and others less than 100 MeV.

3.2.2 Production of hydrogen, deuterium, and helium:
The Universe a few seconds old; Lepton Epoch

After one millisecond, that is, in the middle of the Lepton Epoch, the temperature and density fall to values which permit the survival of only the most stable particles. These include the proton, the neutron (quasi-stable), and the electron. The neutrinos, although very numerous, can be ignored here since they play no role in the events of this epoch, acting only by means of the weak force. At this moment the temperature has fallen to 10 billion K and the corresponding mean photon energy is ~ 1 MeV. These photons cannot influence the "strong nuclear forces" between the nucleons, but they can and do influence the electromagnetic interaction between protons and electrons. This gives rise to the fact that, at this period of the evolution of the Universe, protons and electrons exist separately and are not bound together in a neutral combination forming the hydrogen atom.

The weak force now begins to play an increasingly important role. (We recall that the beta-decay process is governed by the weak force; see Chapter 1.) The three following processes are possible (Figure 3.2):

a) beta-minus decay process (e.g., for free neutron):

$$\text{neutron} \xrightarrow[t_{1/2} = 636 \text{ s}]{\text{beta-minus}} \text{proton} + \text{electron} + \text{anti-neutrino};$$

b) beta-plus decay (possible only in proton-rich atomic nuclei):

$$\text{proton} \xrightarrow{\text{beta-plus decay}} \text{neutron} + \text{positron} + \text{koino-neutrino};$$

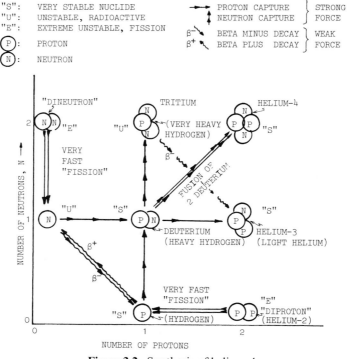

Figure 3.2. Synthesis of helium-4.

3.2 Evolution of the Elementary Particles.

c) electron capture process:

$$\text{proton} + \text{electron} \xrightarrow{\text{electron capture}} \text{neutron} + \text{neutrino}.$$

The strong nuclear force is, however, still of importance and initiates other nuclear processes – the first nucleosynthesis, for example, which is the synthesis of a deuteron with the emission of a gamma quantum:

$$\text{neutron} + \text{proton} \xrightarrow{\text{fusion}} \text{deuteron} + \text{gamma quantum}.$$

The deuteron is the atomic nucleus of heavy hydrogen, deuterium, and consists of one neutron and one proton. Deuterium has a mass number $A = 2$. The atomic number A represents the sum of the number of protons ($Z = 1$) and number of neutrons ($N = 1$). It is easy to see that:

$$\underset{(A)}{\text{mass number}} = \underset{(Z)}{\text{atomic number}} + \underset{(N)}{\text{neutron number}}.$$

Deuterium is beta-stable and does not decay spontaneously. This newly synthesised deuterium can react with further protons or neutrons or even with other deuterium atoms. In the Lepton Epoch the following syntheses are also possible:

a) *synthesis of tritium*:

$$\text{deuterium} + \text{neutron} \xrightarrow{\text{fusion}} \text{tritium} + \text{gamma quantum}.$$

Tritium (T) is the heaviest isotope of hydrogen and is beta-unstable, it decays with a half-life ($t_{1/2}$) of ~ 12 years to the stable helium-3:

$$\text{tritium} \xrightarrow[(t_{1/2} = 12 \text{ years})]{\text{beta-minus decay}} \text{helium-3} + \text{electron} + \text{anti-neutrino}.$$

b) *synthesis of helium-3*:

$$\text{deuterium} + \text{proton} \xrightarrow{\text{fusion}} \text{helium-3} + \text{gamma quantum}.$$

c) *burning of deuterium and synthesis of helium-4*:

$$\text{deuterium} + \text{deuterium} \xrightarrow{\text{fusion}} \text{helium-4} + \text{gamma quantum}.$$

Both isotopes of helium, He-3 and He-4, are very stable. As we have seen, the primordial matter passes through some very important stages during the first second of the life of the Universe. The isotopes of hydrogen (hydrogen, deuterium, and tritium) and the isotopes of helium (He-3 and He-4) have been produced from the free neutrons and protons.

This is a good point at which to make some general observations which are of great importance to the development of Man and his environment and relate to the principal theme of this book:

> Cosmic matter is rich in hydrogen and helium formed by nucleosynthesis during the early phases of the Universe.

3.2.3 The Photon Epoch, from the first minute to the first million years

The Universe is now some hundreds of seconds old. The Lepton Epoch has just come to an end. The Universe is much bigger, with a radius of over 10^{18} m. The density decreases to $\sim 10^3$ kg/m^3, the temperature is 100 million K, and the mean photon energy equals 10 kilo-electron volts (10 keV). Matter consists of the atomic nuclei of hydrogen, deuterium, and helium (He-3 and He-4).

The numerous collisions between these atomic nuclei cannot now result in fusion, since the electrical repulsion force due to the nuclei being positively charged is now much higher than the collision energy. No nuclear reactions are possible. Nuclear synthesis is completely at a standstill. The evolution of matter has come to an end. Forever? No, only while the Photon Epoch lasts.

But another very important process is assumed to take place during this Photon Epoch.

One million years after the Photon Era begins, however, it too comes to an end. We know that at this moment the radius of the Universe is 10^{22} m, corresponding to a distance of approx. one million light years. The density decreases to 10^{-23} kg/m^3, the temperature to ~ 3.5 thousand K, and the mean photon energy to 0.4 electron volts.

At this point the properties of cosmic matter change dramatically. The negatively charged electrons can combine with the positively charged atomic nuclei, resulting in the formation of neutral atoms.

For example, the following simplified reactions can be shown:

$$\text{atomic nucleus of hydrogen} + \text{electron} \xrightarrow{\text{recombination}} \begin{cases} \text{hydrogen} \\ \text{atom} \end{cases}$$

$$\text{atomic nucleus of helium} + 2 \text{ electrons} \xrightarrow{\text{recombination}} \begin{cases} \text{helium} \\ \text{atom} \end{cases}$$

The interaction of the electrically neutral atoms with photons is much less intense than is the case for the charged atomic nucleus and free electrons. Now matter is transparent to photons. In earlier phases, cosmic matter was in the form of a plasma, which is a mixture of free electrons and free atomic nuclei, both of which react intensely with photons; matter was not transparent. We call the period of these early eras together "the Fireball". Now the Fireball has died out and a new era begins – the era of a cold, transparent, and low-density Universe.

3.3 The Beginning of the Present Very Cold Era: The "Stars Era". The Evolution of Galaxies, Stars, and Life

3.3.1 The largest of the cosmic structures: The development of galaxies

The Photon Epoch has just finished and, with it, the primordial Fireball. The Universe is one million years old. The "Stars Era" begins and continues to the present time.

The Universe is cold, the temperature falls below 1000 K, and after one billion (10^9) years reaches ~ 10 K, and today, after 12 billion (12×10^9) years, is only 2.9 K.

During the Fireball, the Universe was probably more or less homogeneous and contained no "structures" with the exception of neutrino clumps. Now, in the cold Star Era, the situation changes significantly. The temperature is low, the kinetic energy of the atoms is low, and the forces of gravitational attraction start to come into play. The attraction produces large cold "clouds" containing gas with a total mass of over 10^{42} kg. These are the protogalaxies, probably made up of mostly neutrinos.

The origin and evolution of the galaxies is still imperfectly understood. However, we know there are approximately 100 billion (10^{11}) galaxies in the Universe.

At least three types of galaxies can be simply classified (Table 3.1):

a) quasars, very young and very distant, with an extremely high luminosity of 10^{39} watts;
b) radio galaxies, nearer to us and with a luminosity of 10^{38} watts; and
c) normal galaxies with a luminosity of 10^{37} watts. These normal galaxies can be subclassified into at least four classes. Our galaxy, the Milky Way, is one such normal, average-size spiral galaxy with a luminosity of 10^{37} watts. A galaxy of this size typically contains approx. 10^{11} stars. A scheme of the Milky Way is shown in Figure 3.3.

3.3.2 The evolution of stars; the nuclear and gravitational reactors

Just after the birth of the galaxies, a large proportion of the diffuse matter condenses in the form of small but dense "drops" containing approximately 10^{55} to 10^{59} atoms, that is, a mass of 10^{28} to 10^{32} kg. These are the stars. Taken together in the whole Universe, the 10^{11} to 10^{12} galaxies that exist contain approx. 10^{23} stars. (It is interesting to note that one mole of atomic hydrogen, that is, one gram, contains approx. 6×10^{23} atoms.)

Table 3.1. Different Types of Galaxies

Type	Spiral galaxy	Irregular galaxy	Giant elliptical galaxy	Quasar	Radio galaxy
Mass (solar masses)	10^{12}	10^9	10^{12}	10^{11} ?	10^{11} ?
Diameter (light-years)	100 000	20 000	150 000	0.01	100 000
Luminosity (Watt)	10^{39}	$<10^{35}$	10^{37}	10^{40}	10^{38}
(Sun's luminosity)	10^{12}	10^9	10^{10}–10^{11} (X-ray luminosity $\sim 10^9$)	10^{14}	Radio wave
Age (gigayears)	~ 10	~ 10	~ 10	Very young ~ 1	Young ~ 5
Colour	Blue (disc) Red (halo)	Very blue	Red	Radio + + blue	Radio
Gas content	$\sim 10\%$	25 %	$<1\%$	?	?
Types of stars	Young (disc) Old (halo)	Young	Old	...	...
Example	Our galaxy „Milky way" (Fig. 3.6)	Magellanic clouds	M 87 Part of virgo cluster	Quasar 3C 273 (the nearest quasar)	Radio galaxy Centaurus A
Distance to the Earth (light-years)	—	$\sim 200\,000$	$>10^6$	1.5×10^9	$\sim 10^8$
Central object	see Fig. 3.3	?	May contain a central black hole with mass of $\sim 10^8\,M_\odot$ and radius of 350 light-years	Black hole?	Black hole?
Rotation period (years)	2.0×10^8	?	?	?	?
Number in Universe	10^{11}	$<10^{11}$	$\sim 10^{11}$	$\sim 10^{6\,a}$)	Small

[a] Catalogue 1980 contains 1500 Quasars.

Galaxies are long-lived, having existed for perhaps ~ 12 billion years (12×10^9 years). However, the individual stars have much shorter lives, on the average one or two billion years, but up to 10 billion years and even longer. The largest stars, with a mass of 10^{32} kg, that is, over 100 times more massive than our own Sun (mass of Sun: $M_\odot = 2 \times 10^{30}$ kg), have a rather short lifetime of one million or as little as one hundred thousand years.

3.3 The Beginning of the Present Very Cold Era: The "Stars Era"

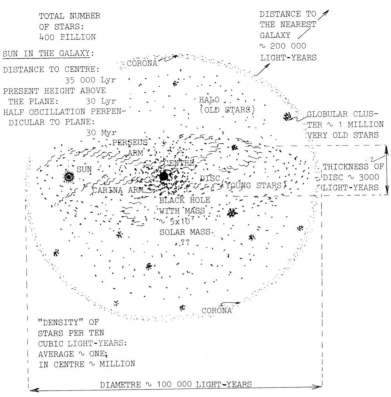

Figure 3.3. The present structure of our galaxy, the Milky Way.

New stars evolve from the "ashes" of the dead stars, and become the second generation.

3.3.3 The protostar evolves from diffuse matter

Due to the attraction of gravitation the old diffuse matter, gas and dust, is drawn together and contracts. The protostar is formed. During this contraction period of its birth only those chemical elements formed in the "Big Bang" nucleosynthesis, that is, mostly hydrogen, helium, and small amounts of lithium and beryllium. During their active life these stars partially burn and transform the hydrogen and helium into elements with a greater mass. The second-generation stars include both materials from the "Big Band" and the products of the burning mechanism of the first-generation stars. Massive second-generation stars also have relatively short lives and expire in a dramatic explosion. From these "ashes" a third generation of stars evolves. Our Sun is probably a star of this type. We can say finally that in one galaxy there can be, at any one time, stars of different ages, of different masses, and of different generations.

It is possible to classify the numerous star types by means of two parameters, both of which can be observed – the temperature of the surface and the amount of energy emitted per unit time, that is, the power or luminosity.

At the beginning of this century, two observers, Russel and Hertzsprung, independently coupled together these two parameters for the then-known stars. Figure 3.4 shows the Hertzsprung–Russel diagram.

Most stars lie on a line going from the upper left-hand corner to the lower right-hand corner, the so-called "Main Sequence". The reason is really trivial. Those stars which live longest come to represent the largest proportion of all stars. These same long-lived stars have obviously similar properties, therefore the relationship between temperature and luminosity is similar. The Hertzsprung–Russel diagram could also be drawn up in the form of a radius and density relationship for the Main Sequence stars.

Those stars having the highest luminosity of approx. 10^{30} watts, that is, ten thousand times more powerful than the Sun, belong either to the Main Sequence stars, the so-called Blue Giants with surface temperatures of thirty thousand kelvins, or to the class of Red Giants with surface temperatures of some thousand kelvin. At the lower level of luminosity there are stars of

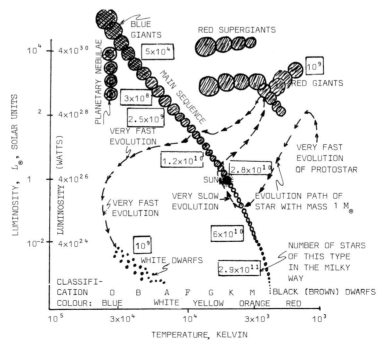

Figure 3.4. Hertzsprung-Russel diagram of stars' evolution.

10^{23} watts, that is, one thousand times less luminous than our Sun. Here the stars with surface temperatures of more than ten thousand kelvin belong to the class of White Dwarfs, and those with surface temperatures of a few thousand kelvin to the Main Sequence stars. The Blue Giants have a mass approx. one hundred times greater than that of the Sun and have very short lives, significantly less than one million years. The White Dwarfs, having a mass similar to that of the Sun, live very much longer, probably tens of billions of years.

It is of value now to discuss the evolution of the stars, to consider their chemical composition, to analyse the way the most important force – gravitation – is transformed into thermal energy. The matter in the protostar is heated and the temperature rises to a few thousand kelvin. Matter becomes ionised, that is, part of the electron population is separated from the atoms. These electrically charged components interact with photons much more readily than the neutral matter. The protostar is now opaque to the photons, which escape from the interior of the protostar only after a longer period of time. This results in an increased heating rate and the temperature rises to 10^7 K. The first nuclear processes become possible. Up to this moment the single source of energy was the gravitational force, but now the phase of nuclear energy release begins. We should note, however, that only protostars with a mass greater than 10^{28} kg can, under the influence of gravitational contraction, heat up to 10^6 K.

The release of nuclear energy begins with reactions involving the light elements produced in the Big Bang, with the exception of the very stable helium-4. The first components of stellar fuel are the nuclides deuterium, lithium, and beryllium. These nuclides "burn" at the relatively low temperature of 10^6 K. Significant amounts of energy are released; the interior temperature rises, as does the surface temperature, reaching more than five thousand kelvin, making it visible. The life span of the protostar is, however, only about 1000 years.

Protostars having a mass of lower than 10^{28} kg, that is, less then 1/100 of the Sun's mass, are unable to reach the million kelvin point because the gravitational contraction is too weak. These objects remain as cold black spheres. The question is, how many such objects exist in a galaxy?

The evolution of a protostar of an average stellar mass results in the origin of a double star, and sometimes even a triple-star system. The multiplicity of solar-type stars is greater than has been previously thought. Each primary star has 1.4 companions on the average, which means that its multiplicity equals 2.4.

Two-thirds of the stars have stellar companions and the remaining one-third may be expected to have close non-stellar companions. Part of the latter are massive objects – the planets (see Chapter 4.3). Even the Sun has been suspected to have a small companion, e.g., a brown dwarf with a mass of ~ 0.01 solar mass and a maximum distance of less than 2.5 light-years.

3.3.4 The longest living stars, those of the Main Sequence

Due to the burning of deuterium, lithium, and beryllium in the interior of these newborn stars and the opacity of the plasma, the temperature rises to $1–1.5 \times 10^7$ K. A new period in the life of the star begins. The most abundant component, hydrogen, begins to react; it is burned and is transformed into helium-4. When we remember that about 93 of every hundred atoms in cosmic matter are hydrogen atoms and that each atom transformed into helium-4 releases approx. 7.6 MeV of energy, we see that a tremendous energy source exists in the stars.

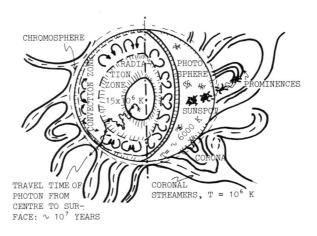

Table 3.2. The Sun Today

Property	Unit	Value
Mass	kg	1.99×10^{30}
Mass, relative	$M_\odot$ (solar mass)	1.00
Radius	metre	6.96×10^8
Radius, relative	$R_\odot$ (solar radius)	1.00
Density, average/centre	kg/m^3	$1.41 \times 10^3 / 135 \times 10^3$
Luminosity	Watt	3.90×10^{26}
Luminosity, relative	$L_\odot$ (solar luminosity)	1.00
Pressure in centre	bar	2×10^8
Temperature in centre	Kelvin	14.6×10^6
Temperature at surface	Kelvin	~ 5785
Energy source	kg/year	Hydrogen burning: 1.92×10^{19}
Energy flux, surface	W/m^2	63×10^6
Specific power, mean	W/kg	1.88×10^{-4}
Age	y	$\sim 4.6 \times 10^9$
Velocity relative to the local stellar matter	m/s	$\sim 20,000$

3.3 The Beginning of the Present Very Cold Era: The "Stars Era"   43

Hydrogen is burned in the star until about one-tenth of this fuel has been exhausted. In a star of average mass, approx. 10^{30} kg and luminosity of 10^{26} watts, the hydrogen reaction lasts a long time, some billion years (Table 3.2). These are the Main Sequence stars. The Sun is a typical Main Sequence star (Table 3.2). The more massive stars burn much more intensely and achieve a much higher luminosity, therefore their life is shorter. These are the Blue Giants.

The transport of energy from the star's interior to the surface and into cold cosmic space is carried out by photons.

Most stars in the Galaxy belong to the Main Sequence.

3.3.5 Red Giants: The cold stars with the hot interiors

When one-tenth of the hydrogen fuel has been consumed, the star enters another phase. The helium produced in the process accumulates in the interior of the star and prevents further "burning" of the hydrogen. Instead of evolving further free energy, the centre of the star begins to contract under the influence of gravity. The contraction causes the interior temperature to rise and at the same time the temperature of the outer shell, containing more hydrogen and less helium, rises. The outer shell expands and then cools. The star has now expanded and has a rather cold outer shell and a dense interior rich in helium. At this stage of its evolution the star is a "Red Giant" (Figure 3.5). The Red Giant stage is rather short, some millions of years.

TYPE OF STAR	UNIT	RED GIANT	SUPERNOVA	WHITE DWARF	NEUTRON STAR
STRUCTURE OF STAR		$3He^{12}C$ $4H \rightarrow He$	VELOCITY $\sim 10^6$ m/s EXPLOSION He-BURNING IMPLOSION HEAVY ELEMENTS SYNTHESIS	He OR Fe	X-RAYS EMISSION VERY STRONG MAGNETIC FIELD
MASS (IN SOLAR MASSES)	M_o	$\sim$ 1-2	4-8	< 1.4	> 2
RADIUS	m	10^{11}	>> 10^{12} INCREASING	10^7	$\lesssim 1.1 \times 10^4$
DENSITY (CENTRAL)	Mg/m	$\sim 10^2$	>> 10^6 INCREASING IN CENTRE	$10^6 - 10^8$	$5 \times 10^{14} - 2 \times 10^{15}$
TEMPERATURE (CENTRAL)	K	< 10^9	3×10^9	< 10^7	$\sim 10^9$ (?)
LUMINOSITY (IN SOLAR UNITS)	L_o	> 100	2×10^8	$\sim$ 0.0001	10^{-6} IN RADIOWAVES
LIFE	YEARS	$10^7 - 10^8$	DAYS (HALF-LIFE $\sim$ 55 DAYS)	> 10^9	> 10^7 (?)
STRUCTURE	---	TWO ZONES	MULTIPLE ZONES	ONE ZONE	MULTIPLE ZONES
ESTIMATED NUMBER IN MILKY WAY	---	HUNDRED THOUSAND?	ONE EVERY 35 YEARS	TENS OF MILLIONS	$\sim$ MILLION

Figure 3.5. Most important types of stars (excluding Main Sequences stars, see Table 3.2).

3.3.6 Evolution towards hot dense stars

Stars having a mass large enough for the initial hydrogen reaction evolve now into a fairly active stage. As the temperature rises due to the contraction of the interior, the helium begins to react, forming carbon, oxygen, and neon. After the helium burning has begun the fusion reactions become more and more intense. The centre of the star becomes still hotter than the surrounding zone, where hydrogen fusion is occurring. This can give rise to an explosion of the star's interior which, however, is intercepted by the outer shell. This event is called "helium flash". The star, having experienced such an explosion of its central region, becomes hotter and smaller. A new interior is formed, having two zones. The inner region contains the medium heavy elements oxygen, carbon and neon, while the outer shell is made up of helium. As before, this interior is surrounded by the outer hydrogen burning zone.

If the mass of the star is large enough, the following process begins. When the temperature reaches some billions of kelvin, not only helium-4 but also the relatively stable nuclides such as carbon, oxygen, and even silicon begin to fuse together. The products of this nuclear fusion are the nuclides iron, nickel, and other metals. These different sources of nuclear energy are ignited in ever decreasing time spans. With a large enough mass, the star can experience, in

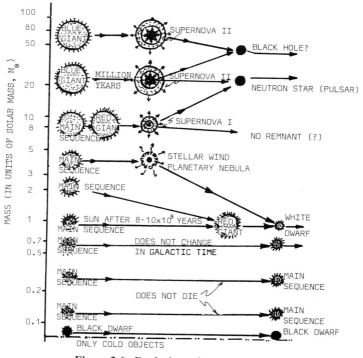

Figure 3.6. Evolution of stars.

3.3 The Beginning of the Present Very Cold Era: The "Stars Era"

more advanced phases of its development, several more fusion processes at the same time. The star can be compared to an onion made up of many skins, with a different nuclear process occurring in each skin (Figure 3.5). In the outer zone the fusion of hydrogen always occurs, while in the innermost zone the fusion of the heaviest elements are ignited at that stage.

Depending on the mass of the star, the point is eventually reached whereby the energy release due to gravitational attraction is unable to ignite any fresh nuclear reactions. Then begins a slow cool-down period, lasting many billions of years. The star contracts; its energy source is gravitational and energy production due to nuclear reactions falls off. The resulting contraction lasts until repulsions between the atomic nuclei are able to resist any further contraction. The star becomes a "White Dwarf" (Figure 3.6). The White Dwarf has a mass less than 1.4 times the Sun's mass, but a volume similar to that of a planet. It has, therefore, a very high density – about one million times that of water. The surface remains relatively hot for a long time, since due to the reduced surface area the remaining energy can only slowly leak away. They cool down slowly and finally become invisible (Figure 3.6).

3.3.7 Explosion of a supernova: The most spectacular event in a galaxy

Stars with a mass 6–10 times that of the Sun have a very dramatic life.

A massive star which has achieved the temperature of billions of kelvin and is then cooled by neutrino emission undergoes a collapse of its inner zones. The collapse into the centre releases a very large amount of energy in a very short period. Within seconds we can speak of an implosion and then, as the outer shell begins to expand, we see an explosion. The energy release in the first moments due to gravitational collapse rises to a level of some 10^{35} watts, that is, a luminosity equal to the total luminosity of the average galaxy. From one star! This enormous energy release decays with a half-life of approx. 55 days. We have witnessed a supernova (Figures 3.5 and 3.6). The supernova dissipates a large amount of matter into interstellar space. Mankind has registered about 14 supernova explosions in our Galaxy in the last twenty centuries, the last one in 1604, and in other galaxies approx. 400 supernovae explosions.

We have now realised that stellar deaths (in the form of supernovae) and stellar births are intimately connected.

3.3.8 Extremely dense stars: Neutron stars (pulsars) and black holes

The remains of a supernova explosion consist of the imploded internal shells. The gravitational collapse causes very high pressures, leading to enormous density increases. Protons and electrons are forced together to form neutrons, making a neutron star. Chemical elements no longer exist – all the matter has been "neutronised". It is probable that neutron stars emit radio waves in short pulses, hence the name pulsars (Figure 3.5). At present we know thousands of pulsars, mostly near the galactic centre. The total number in the galaxy has been recently estimated as one to two million. The density of a neutron star is the same as the density of nucleons, that is, 10^{17} kg/m^3. The diameter of the star is only on the order of tens of kilometers. The existence of a black-hole star is still not established.

3.4 The Burning of Hydrogen – Nucleosynthesis in the Stars

3.4.1 Deuterium: The fuel of protostars

It has been shown how the protostar achieves a temperature of approx. 10^6 K in its interior due to gravitational contraction. This corresponds to the phase where nuclear fusion (called, for simplicity, "burning") begins. Deuterium begins to burn first. Why not hydrogen? Hydrogen is some thousand times more abundant than deuterium, and it releases more energy per unit mass. Why is deuterium the first to burn? The reason is as follows, and we must emphasise that this difference between the burning of hydrogen and deuterium results in events of the highest importance for the evolution of the Universe, not least for the evolution of mankind. Here is the simple explanation:

a) Hydrogen burns to form helium-4:

$$4 \text{ atoms hydrogen} \xrightarrow{\text{slow}} \text{helium-4} + 2 \text{ positrons} + 2 \text{ neutrinos} + \text{energy}.$$

b) Deuterium burns to form helium-4:

$$2 \text{ atoms deuterium} \downarrow \xrightarrow{\text{fast}} \text{helium-4} + \text{energy}.$$

As we can see, hydrogen burning produces neutrinos and this is clear evidence that this process, among others, is controlled by the "weak force". Oppositely, the fusion of deuterium releases no neutrinos (anti- or koino-), since it is controlled by the "strong force" (Chapter 1.2). We have seen that the "weak force" is not only weak but initiates reactions only slowly. The "strong force", on the other hand, is associated with a much higher reaction rate. Thus the hydrogen burning proceeds at a much slower rate. At a temperature of only 10^6 K, only deuterium burning, controlled by the strong force, is able to achieve a sufficiently high intensity. Other reactions occur with similar intensity.

Deuterium, lithium, and beryllium burn so rapidly that all of these nuclides which experience stellar burning are completely exhausted. Another source must eventually be found for these light elements outside the environment of a star to explain their continuing existence in cold cosmic matter, including the matter of our planet.

3.4.2 The slow burning of hydrogen

With the burning of deuterium, lithium, and beryllium complete and at a temperature of ~ 15 million K, the longest phase, that of the Main Sequence Star, begins. The phase is characterised by the burning of hydrogen.

The nucleosynthesis of helium-4 may be divided into the three steps, in which only two particles react at one time. We make the obvious assumption that a mutual collision of three, not to mention four, particles is much less likely.

The burning of hydrogen, which results in the synthesis of helium-4, is without doubt one of the most important and most powerful of the processes governed by the elementary weak force at this point in the development of the Universe.

We have not yet mentioned the energy balance of this three-step process, called the "pp-process" (proton-proton-process) (Figure 3.7(a)):

$$4\,H \rightarrow {}^4He + 2\,e^+ + 2\,v + Q; \quad Q = 26.2\,\text{MeV} = 6.4 \times 10^{14}\,\text{J/kg}$$

$$4\,\text{hydrogen} \rightarrow \text{helium} + 2\,\text{positrons} + 2\,\text{neutrinos} + \text{energy}.$$

We have seen that hydrogen burning releases approx. 6.4×10^{14} J/kg and that the process is governed by the weak force (step a), and therefore proceeds slowly. Let us assume a rate of consumption for a kilogram of hydrogen, containing 6×10^{26} atoms, of only 6×10^8 atoms per second. We know that hydrogen burning stops when 10 percent of this fuel is exhausted. We can now calculate the time needed to consume all the available hydrogen in a star:

$$\frac{(0.1) \times (6 \times 10^{26}\,\text{atoms/kg})}{(6 \times 10^8\,\text{atoms/s} \cdot \text{kg})} = 10^{17}\,\text{s} = 3 \times 10^9 \,\text{years}.$$

The specific power given by this rate of burning equals:

$$6 \times 10^8 \left(\frac{\text{atoms}}{\text{s} \cdot \text{kg}}\right) \times \frac{26.2}{4} \left(\frac{\text{MeV}}{\text{atom H}}\right) \times 1.6 \times 10^{-19} \left(\frac{\text{J}}{\text{eV}}\right) \cong 6.3 \times 10^{-4}\,\frac{\text{watt}}{\text{kg}}.$$

In the case arbitrarily chosen here, where the rate of burning for 10^{18} atoms (that is 1.67 µg) is one atom per second, the specific power is 0.65 mW/kg and the period of burning is about 3 billion years.

Our Sun, however, has a specific power of only 0.188 mW/kg and, therefore, the period of hydrogen burning can be as long as 9 billion years. It is, in fact, only about 5 billion years old, and thus still has some time to go.

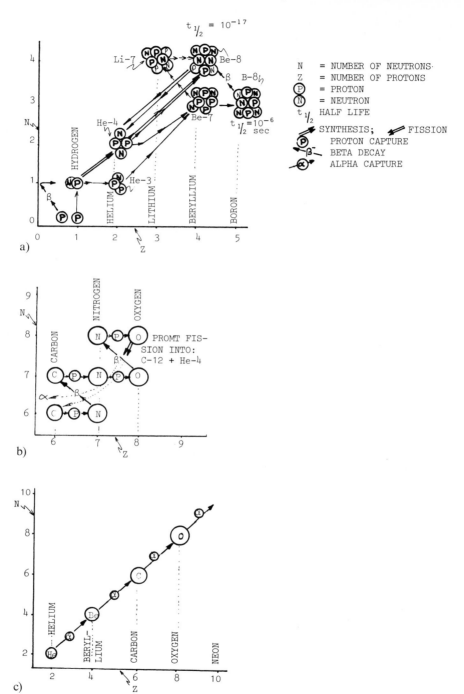

Figure 3.7. (a) Hydrogen burning: "pp-process". **(b)** Catalytic burning of hydrogen: "CNO-process". **(c)** The burning of helium, the "three-alpha-process" (3α-process).

3.4.3 The burning of hydrogen in a catalytic cycle assisted by carbon, nitrogen, and oxygen

The direct burning of hydrogen by means of the synthesis of deuterium – the pp-process – is not the only one possible. There is another in which the elements carbon, nitrogen, and oxygen play the role of catalysts, that is, increase the rate of the reaction (Figure 3.7 (b)). This process can only take place, of course, in those stars which contain enough carbon.

3.5 Helium also Burns, but under More Extreme Conditions

3.5.1 Production of carbon from the burning of helium

Helium-4 is an end-product of hydrogen burning. It is a form of nuclear ash. Helium-4 is significantly poorer in energy than hydrogen. During its synthesis from hydrogen about 26 MeV have already been released. Is this, then, the end of the nuclear processes as an energy source? The answer is no. There are many elements of middle mass in which nucleons are still more firmly bonded than helium-4. Consequently, helium-4 can burn to form these elements and release further energy. As a fuel, however, the burning takes place under more extreme conditions. It can burn only at high temperatures over 200×10^6 K and only when the density is high enough (10^6 kg/m^3). These conditions can be reached in the inner regions of the Red Giants (Figure 3.5).

Even here the burning of helium-4 is not without problems. Some of them are discussed below; the simplest helium-4 reaction is given first (Figure 3.7 (c)):

a) Helium-4 fuses with another helium-4:

$$\text{helium-4} + \text{helium-4} \xrightarrow[\text{(strong force)}]{\text{fusion (fast)}} \text{beryllium-8} - \text{energy}.$$

Note that the energy balance is negative. Energy is not released, but on the contrary must be taken from another source:

b) Beryllium-8 is very unstable and fissions:

$$\text{beryllium-8} \xrightarrow[\text{(strong force)}]{\text{fission (fast)}} \text{helium-4} + \text{helium-4} + \text{energy}.$$

Note that the energy balance is positive in this case. Experimental measurements of the half-life of beryllium-8 give a value of only 10^{-16} second. This is not very encouraging but there is a way around the problem.

In the centre of a Red Giant, where the density can be as high as 10^6 kg/m^3 and the temperature over 200×10^6 K, there is, in fact, a chance that the unstable beryllium-8 atom will react with a further helium-4. The new situation can be written as follows.

c) Beryllium-8 reacts with helium-4 before it has a chance to decay through fission:

$$\text{beryllium-8} + \text{helium-4} \xrightarrow[\text{(strong force)}]{\text{fusion (very fast)}} \text{carbon-12} + \text{energy}.$$

d) The balance of reactions a, b, c can be written:

$$3\,\text{helium-4} \xrightarrow[\text{(strong force)}]{\text{fusion, fast}} \text{carbon-12} + \text{energy (7.24 MeV)}.$$

3.5.2 A very vital step: The production of oxygen

Helium burning, and thus the synthesis of carbon-12, is only the first step in a long chain of nucleosynthesis taking place in the heart of a Red Giant. Some of the steps are as follows.

a) Synthesis of oxygen:

$$\text{helium-4} + \text{carbon-12} \xrightarrow[\text{fast}]{\text{fusion}} \text{oxygen-16} + \text{energy (7.1 MeV)}.$$

b) Synthesis of neon-20:

$$\text{helium-4} + \text{oxygen-16} \xrightarrow[\text{fast}]{\text{fusion}} \text{neon-20} + \text{energy (4.73 MeV)}.$$

c) Synthesis of magnesium-24:

$$\text{helium-4} + \text{neon-20} \xrightarrow[\text{fast}]{\text{fusion}} \text{magnesium-24} + \text{energy (0.31 MeV)}.$$

These reactions produce the relatively abundant elements carbon, oxygen, neon, magnesium, and silicon.

In this group we have the elements essential to life (Chaper 4) and the most important components of the Earth's crust (Chapter 5).

3.6 Carbon, Oxygen, and Other Elements of Medium Mass Burn in a Flash

3.6.1 Energy production and energy required for nucleosynthesis

During the relatively slow evolution of the stars, a small but nevertheless very important step has been covered. Starting from hydrogen and deuterium, helium, carbon, oxygen, neon, and nitrogen have been synthesized. Only another 80 elements have yet to be produced.

3.6.2 Iron, the nuclear ash

The elements of middle mass from magnesium (atomic number $Z = 12$) to iron (atomic number $Z = 26$) and nickel (atomic number $Z = 28$) are synthesised via explosive burning. As the most stable nuclides, for example, $^{56}_{26}\text{Fe}_{30}$ (iron, with atomic mass 56: 26 protons and 30 neutrons), are approached, the possibility of an exothermic reaction nears its limit. No further reactions are possible. For synthesis of other, heavier elements a supply of energy from outside is required. Additional production by means of strong interactions cannot occur, since the inherent nuclear forces are fully exhausted. The limit is marked by the boundary zone, iron. The concept of an iron curtain is rather appropriate: The special role of iron in the story of the Universe cannot be overemphasised. Wherever material exists at atomic densities, iron is the most stable of the nuclei.

At this point, three important observations should be made:

a) The net energy production during the first step of the synthesis is large:

4 hydrogen atom → helium-4 atom + free energy (26 MeV).

This 26.7 MeV per 4 nucleons gives approx. 6.6 MeV per nucleon for a jump from atomic number $Z = 1$ to atomic number of $Z = 2$,

b) From the form of the curve of binding energy (see Figure 1.7) it can be seen that energy is released from each reaction up to the synthesis of iron-56, $^{56}_{26}\text{Fe}$, the most stable of all isotopes and the ash of all exothermic nuclear processes.

For the balance of energy produced in the formation of iron, starting from oxygen-16, the following equation can be written:

$3\frac{1}{2}$ atoms of oxygen-16 → 1 atom of iron-56 + energy (1 MeV).

The total net energy gain is 1.0 MeV, or approx. 0.02 MeV per nucleon, for a jump from atomic number $Z = 8$ to atomic number $Z = 26$.

c) Further steps in which the atoms from iron $Z = 26$ to uranium $Z = 92$ are synthesised require a consumption rather than production of energy:

$4\frac{1}{4}$ atoms of iron-56 + energy (270 MeV) → 1 atom of uranium-238,

and so the energy consumption is approx. 1.2 MeV per nucleon.

3.7 The Synthesis of Heavy Elements: The Need for an External Energy Source

3.7.1 How can uranium be synthesised?

Now we want to learn more about the next steps in nucleosynthesis. Starting from iron, Nature must synthesise the remaining elements up to uranium (atomic number $Z = 92$) or even plutonium ($Z = 94$).

Thus the problem of nucleosynthesis of the heavy elements is governed by the following requirements:

a) a source of free energy of about 1.2 MeV per nucleon when starting with iron-56 (it cannot be a nuclear source, therefore only gravitational forces can be utilised, but now with an entirely different order of magnitude);
b) a source of short-lived neutrons with a high enough intensity;
c) a rich source of iron atoms;
d) the whole process must not have occurred in the too distant past. This limitation relates to the fact that uranium is rather unstable and undergoes radioactive decay. Less than 6 billion years ago each isotope was synthesised in equal quantities, but the ^{235}U decayed approx. 6.3 times faster than the ^{238}U, leading to the present relative abundance of 1:140. This process of nucleosynthesis must have occurred in a cosmic object not much older than our Sun or the Earth, having an age of approx. 5×10^9 years. In addition, the object must have existed in this Galaxy, as the transport of material from another galaxy is virtually impossible, or at least takes too long for the uranium-235 to have survived.

Now comes a vital question: does a cosmic object exist with the following properties?

a) a large but short-lived source of gravitational energy, e.g., a massive object which in a relatively short time shrinks considerably;
b) a high initial iron content;
c) a large, intense, but short-lived source of neutrons.

This object which we are seeking should have triggered the formation of thorium and uranium some 6×10^9 years ago. Can one still observe such objects today? The answer is yes. They are almost certainly the supernovae, the most spectacular events in all galaxies, ours as well as neighbouring or even distant galaxies.

3.7.2 The "s-process", the slow-process of neutron capture

Before we get involved with the nucleosynthesis of thorium and uranium we should look at the synthesis of the elements between iron ($Z = 26$) and lead ($Z = 82$). The heaviest of the stable nuclei, $^{208}_{82}\text{Pb}_{126}$, is double magic: magic number of protons $Z = 82$, magic number of neutrons $N = 126$.

The most important features of this nucleosynthesis are:

a) the ratio of neutrons to protons increases, from approx. 1.10 for iron $^{56}_{26}\text{Fe}_{30}$ to 1.54 for lead $^{208}_{82}\text{Pb}_{126}$;
b) the binding energy decreases (Figure 1.7). From the stable iron the less stable lead is produced. From the nuclear ashes a nucleus containing energy is formed;

3.7 The Synthesis of Heavy Elements: The Need for an External Energy Source

c) the cosmic abundance of all elements between iron and arsenic ($Z=33$) and lead ($Z=82$) is approx. a million times smaller than the abundance of iron (Figure 2.5).

We are discussing here a process in which the neutrons cause a nuclear transformation, the so-called slow process. We must remember that neutrons in the free condition are short-lived, with average lifetimes of approx. 1000 seconds (half-life $\cong 10$ min $\cong 636$ seconds). In order to accept that this is an effective neutron capture process taking place in a star we must demonstrate the existence of a possible long-lived source of neutrons. Such a long-lived source is available in the case of those nuclides having an excess of neutrons over the required stable number, that is, an excess compared to the magic number of neutrons. A typical example is oxygen-17, $^{17}_{8}O_9$, with 8 protons and 9 neutrons.

3.7.3 The "r-process", the rapid-process of neutron capture

The slow process of neutron capture produces only specific nuclides between arsenic ($Z=33$) and bismuth ($Z=83$) with only a small difference in the neutron/proton ratio, typical of the elements of middle mass.

There still remain some nuclides which are beta-stable but which do not figure in the "s-process".

For the synthesis of these nuclides processes other than the s-process come into play. Apparently at least two other separate processes must occur. Let us begin with the synthesis of the neutron-rich and the neutron-balanced nuclides and take one or two typical examples from each case.

a) Synthesis of neutron-rich nuclides: this runs as follows. A nuclide is produced in the s-process, in a very intensive neutron flux such as occurs in the outer region of a supernova, and captures two neutrons in a very short space of time, within, say, a tenth of a second. Take as an example the following reaction:

cerium-140 + neutron $\xrightarrow{\text{capture}}$ cerium-141 $\xrightarrow{\beta^- \text{-decay}}$ praseodymium-141.

Now in a very intense neutron flux it can happen that a second neutron capture occurs before the cerium-141 can decay spontaneously to praseodymium-141:

cerium-141 + neutron $\xrightarrow{\text{capture}}$ cerium-142 + gamma quantum.

b) The neutron-balanced but very heavy nuclides such as thorium-232, uranium-235, and uranium-238 are formed in similar processes, where over a short period of time an intense neutron source is available, that is, in the same supernova.

To explain the formation of uranium-235 in this way a typical chain of events, with capture followed by spontaneous beta-decay, is:

$$\text{lead-208} + 27 \text{ neutrons} \xrightarrow{\text{fast capture}} \text{lead-235} + \text{gamma quanta}.$$

All of these neutron capture events must take place in a very short space of time, of the order of a thousandth of a second. The newly created lead-235, however, is very unstable, and within a similar space of time decays by beta emission:

$$\text{lead-235} \xrightarrow{\text{beta-decay}} \text{bismuth-235}.$$

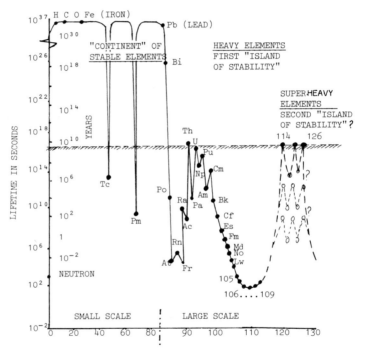

Z	SYMBOL	ELEMENT	Z	SYMBOL	ELEMENT	Z	SYMBOL	ELEMENT
43	Tc	TECHNETIUM	90	Th	THORIUM	100	Fm	FERMIUM
61	Pm	PROMETHIUM	92	U	URANIUM	101	Md	MENDELEVIUM
82	Pb	LEAD	93	Np	NEPTUNIUM	102	Mb	NOBELIUM
83	Bi	BISMUTH	94	Pu	PLUTONIUM	103	Lr	LAWRENCIUM
84	Po	POLONIUM	95	AM	AMERICIUM	104	Rf	RUTHERFORDIUM
85	At	ASTATIUM	96	Cm	CURIUM	105	Ha	HAHNIUM
86	Rn	RADON	97	Bk	BERKELIUM	106	?	?
87	Fr	FRANCIUM	98	Cf	CALIFORNIUM	107	?	?
88	Ra	RADIUM	99	Es	EINSTEINIUM	108	?	?
89	Ac	ACTINIUM						

Figure 3.8. The curve of the lifetime of the most stable isotopes of all elements – age of the earth.

3.8 Cosmic Rays – A Strange Form of Matter

In short, the whole chain looks like this (β^- for beta-minus decay, Bi for bismuth, Po for polonium, etc.):

$$^{235}_{82}\text{Pb} \xrightarrow[\text{extremely fast}]{\beta^-} {}^{235}_{83}\text{Bi} \xrightarrow[\text{extremely fast}]{\beta^-} {}^{235}_{84}\text{Po} \xrightarrow[\text{very fast}]{\beta^-} {}^{235}_{85}\text{At} \xrightarrow[\text{very fast}]{\beta^-}$$

$$^{235}_{86}\text{Rn} \xrightarrow[\text{very fast}]{\beta^-} {}^{235}_{87}\text{Fr} \xrightarrow[\text{fast}]{\beta^-} {}^{235}_{88}\text{Ra} \xrightarrow[\text{fast}]{\beta^-} {}^{235}_{89}\text{Ac} \xrightarrow[\text{fast}]{\beta^-} {}^{235}_{90}\text{Th} \xrightarrow[\text{(5 minutes)}]{\beta^-}$$

$$^{235}_{91}\text{Pa} \xrightarrow[\text{(42.4 minutes)}]{\beta^-} {}^{235}_{92}\text{U (beta-stable)}.$$

The synthesis of uranium-235 is a product of the explosion of a supernova. Other products of supernovae are thorium-232, uranium-238, and even plutonium-244 (Figure 3.8).

3.8 Cosmic Rays – A Strange Form of Matter

We know that a significant amount of cosmic material has the characteristic of motion at a high velocity. Such nuclei can quite often reach energies of 1000 MeV, equivalent to the rest mass of the protons (~ 931 MeV). Thus the kinetic energy of these particles equals the "internal" energy-mass equivalent

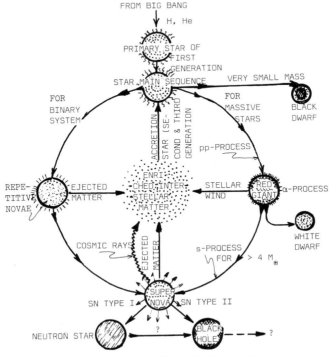

Figure 3.9. Recycling of stellar matter.

of the nucleus itself. However, extremely energetic cosmic radiation can reach 10^{19} eV, in other words 10 billion times as energetic as the rest mass of a proton. Obviously the velocity of these cosmic particles is very close to the limiting velocity – that of light in a vacuum.

Cosmic rays have a complex chemical make-up, extending from hydrogen and helium via the middle elements to uranium and even plutonium. We should not forget that the cosmic rays themselves are more than likely a product of supernova explosions.

The cosmic rays moving between stars and/or galaxies from time to time hit the interstellar or intergalactic material and cause a fission of the nuclei. The nucleosynthesis of light elements in this way is called the X-process, which is also indicative of the uncertainty which exists as to whether elements are, in fact, formed in this way (Figure 3.9).

3.9 What Are the Conclusions for the Future of Mankind?

This chapter dealt with the following problems:

- the evolution of the chemical elements in the Big Bang and in the interiors of stars;
- the relative cosmic abundance of the elements and the important role of hydrogen, helium, oxygen, carbon, and nitrogen in the Universe;
- the history of the evolution of stars and the long life of stars like the Sun;
- the stability and instability of different elements;
- the possibility of extracting energy from the light elements by fusion and the heavy elements by fission.

From this the following conclusions may be drawn for mankind's future development:

- The Earth contains a generous amount of all the stable and quasi-stable elements present in the Universe.
- The abundance of elements in the Sun and on the primitive Earth was suited to the spontaneous creation and evolution of living beings.
- Even the rather rare elements, those especially suited to the extraction of energy, using nuclear processes (deuterium, lithium, uranium, thorium) are present on the Earth in concentrations suitable for the development of advanced technologies.

Again, these are only physical preconditions for mankind's further evolution. The real choice and opportunity lie with Man himself.

CHAPTER 4
Chemical Evolution and the Evolution of Life: The Cosmic Phenomena

It is often said that all conditions for the first production of a living organism are present, which would ever have been present. But if (and oh, what a big if) we could conceive in some warm little pond, with all sorts of ammonia and phosphoric salts, light, heat, electricity etc., present, that a protein compound was chemically formed ready to undergo still more complex changes, at the present day such matter would be instantly devoured or absorbed, which would not have been the case before living creatures were formed.

Ch. Darwin
(1809–1882)

Der allgemeine Lebenskampf des Lebewesens ist
- nicht ein Kampf um die Grundstoffe,
- auch nicht einer um Energie (die in Form von Wärme in jedem Körper reichlich vorhanden ist, leider aber nicht in andere Energieformen umgewandelt werden kann),
- sondern ein Kampf um die Entropie, die durch Energieübertragung von der heissen Sonne auf die kalte Erde verfügbar wird.

L. Boltzmann
(1844–1906)

Life feeds on negative entropy.
E. Schrödinger
(1887–1961)

4.1 Chemical Evolution: Another Phase in the Evolution of Matter

4.1.1 Special case of the electromagnetic force: The chemical force

Up to now we have concentrated on the nuclear processes and the impact of gravitation on cosmic matter. This was necessary because approx. 90% of cosmic matter exists in stars, where the internal temperature reaches tens or hundreds of millions of degrees. In these conditions the kinetic energy of the particles reaches more than a thousand electron volts. Thus the nuclear processes play the dominant role, with all other processes representing a lower energy level.

What are the other processes taking place in matter existing at a lower temperature? Here we find the electromagnetic interactions of atomic nuclei and electrons giving rise to atoms and the electromagnetic interactions between atoms resulting in molecules. We are dealing with chemical relationships.

4.1.2 The actors in the chemical play

Now comes the critical questions: What are the chemical processes relevant to our discussion? What elements must be considered?

It is rather difficult to give a simple picture of the average possible chemical reactions occurring in typical cosmic objects. One reason is that these areas can contain some or all of the ninety elements and the number of possible molecules is on the order of magnitude of tens of millions.

There is one possible way of obtaining this information. It is a shortcut, therefore extremely simplified. For this we take the most general characteristic of the chemical elements, electronegativity, which has for the most electronegative element a value of 4 electron volts (4 eV) and for the least electronegative, ~ 0.5 eV (Figure 4.1).

The potential for chemical reactions to occur is greatest for those elements with the largest differences in electronegativity. Therefore, we can expect the most stable compounds to be formed from the most electronegative elements, fluorine (F) and oxygen (O), on the one hand, and the least electronegative elements, the alkali metals lithium (Li), sodium (Na), potassium (K) and the alkaline earth metals, beryllium (Be), magnesium (Mg), and calcium (Ca), on the other. It is indeed the case that the most stable and most common compounds are the fluorides and oxides of these metallic elements.

The probability of a chemical reaction, however, is influenced not only by the chemical properties of the components, in the above case by the electronegativity, but also by the likelihood of a mutual collision, given by the number of the atoms in question in a given volume. Therefore, a better

4.1 Chemical Evolution: Another Phase in the Evolution of Matter

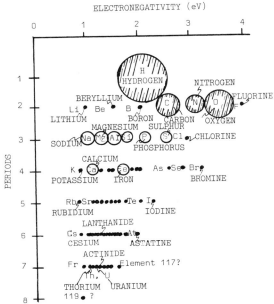

Figure 4.1. The electronegativity of chemical elements and their relative cosmic abundances (not to scale).

understanding of the process of chemical evolution must take into account cosmic abundances. From Figure 4.2 it can be seen that the most abundant elements have an electronegativity greater than 2. The only important exceptions are the elements fluorine and chlorine, with low abundances. In short, we can say the most abundant elements, arranged in decreasing order of abundance, are the electronegative elements hydrogen, oxygen, carbon, nitrogen, sulphur, and phosphorus (Figure 4.2).

This simple information allows us to build up a first broad picture of the chemical evolutionary currents in cosmic matter.

Let us begin with the most electronegative element, fluorine, whose cosmic abundance is approx. 6000 times smaller than that of oxygen, the next most electronegative element. What is more important, however, is that the elements which we expect to combine most readily with fluorine due to their position at the other end of the scale of electronegativity – silicon, magnesium, alkali, and alkaline earth metals – have a much higher abundance (in relative units $\sim 3 \times 10^6$) than fluorine itself ($\sim 10^4$). In addition, the abundances of the chemically similar elements chlorine, bromine, and iodine are so small that we can ignore them. Therefore, in cold cosmic matter all fluorine exists only in the form of the fluoride of one or other of the above metallic elements. But only a small proportion of these metallic elements are bound to fluorine, leaving the larger proportion to react with the next most electronegative element, oxygen, to form oxides.

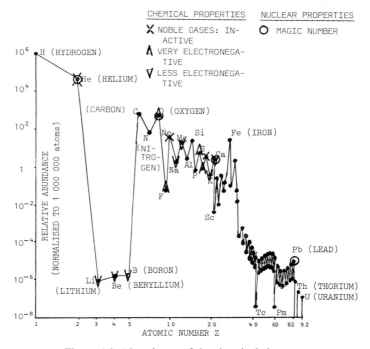

Figure 4.2. Abundance of the chemical elements.

We can see that the relative abundance of oxygen, 2.4×10^7 in relative units, is much greater than all the less electronegative elements including all metals and silicon. Therefore, only a quarter of the available oxygen is bound up with these elements, resulting in nonvolatile oxides of silicon, iron, magnesium, aluminium, sodium, etc.

What happens to the remaining unbound oxygen? The next most likely partner is, of course, hydrogen, the most abundant element in the Universe, with an electronegativity of 2. Carbon also is a possible partner. Nitrogen is not, since its electronegativity is close to that of oxygen.

The simple conclusion is that oxygen reacts with hydrogen and carbon form to oxides of hydrogen, that is, water, and oxides of carbon – carbon monoxide and carbon dioxide.

The chemical reaction between oxygen and hydrogen produces not only water (H_2O) but also hydroxyl (OH). Observation confirms that the hydrogen–oxygen compounds are the most abundant compounds in the Universe. Water is the chief actor on the cosmic chemical stage and, not surprisingly, is coupled with the most dramatic event on this stage – the drama of life.

Using more detailed chemical knowledge we find that not all carbon will bond with oxygen when an element such as hydrogen, showing a higher

4.1 Chemical Evolution: Another Phase in the Evolution of Matter

chemical activity, is present. Quite a large proportion of carbon reacts with hydrogen, resulting in hydrocarbons, initially the simplest, methane (CH_4).

Regarding the other abundant elements – nitrogen, sulphur, and phosphorus – all of these react with hydrogen, forming the following compounds:

- hydrogen nitride, called ammonia: NH_3
- hydrogen sulphide: SH_2
- hydrogen phosphide, called phosphine: PH_3.

Although the abundance of these volatile compounds is much lower than that of water they nevertheless play an important role in the formation of life. This is particularly true of the reactions between these hydrogen compounds and water:

- Methane, CH_4, is almost insoluble in water and does not react with it.
- Ammonia, NH_3, is very soluble in water; its reaction produces ammonium hydroxide, NH_4OH, which has alkaline properties.
- Hydrogen sulphide, H_2S, is soluble in water, stable, and reacts as a weak acid.
- Phosphine, PH_3, is unstable in water and decomposes, producing phosphoric acid, in which the phosphorus atoms are directly and strongly bound by oxygen atoms.

With these simple materials a surprising richness of chemical activity is achieved. This is no less true for living organisms themselves, which also depend on these compounds.

Returning to the chemical evolution of the less electronegative elements, we have seen that they combine most frequently with oxygen. The resulting oxides have different chemical properties, shown most clearly with their reactions with the other most abundant oxide, water:

- silicon oxide, SiO_2: not soluble in water; reacts with other metallic oxides to give very stable, nonvolatile, insoluble compounds, for example, magnesium silicate $MgO.SiO_2$ (or $MgSiO_3$), or the more complex magnesium aluminosilicate $MgO.Al_2O_3.2SiO_2$, that is, $MgAl_2Si_2O_8$;
- aluminium oxide, Al_2O_3: stable, not soluble in water, showing both acidic and alkaline properties;
- iron oxide, FeO: stable, insoluble;
- magnesium oxide, MgO: only slightly soluble with weak alkaline properties;
- sodium oxide, Na_2O: very soluble, with strong alkaline properties.

This very simplified short review of the chemical properties of the most abundant elements can be summarized as follows.

In the colder interstellar cosmic space there are two types of chemical compounds:

- volatile, not very stable and, therefore, very active compounds containing mostly hydrogen and electronegative elements: water (H_2O), methane (CH_4), ammonia (NH_3), and hydrogen sulphide (H_2S);
- nonvolatile, stable, nonreactive compounds containing oxygen: silicon oxide (SiO_2), iron oxide (FeO), aluminium oxide (Al_2O_3), and magnesium oxide (MgO). To this class must also be added elementary carbon (C) in the form of graphite.

4.2 Chemical Synthesis Occurs in Cosmic Space

4.2.1 Interstellar gas contains very many, often very complex compounds

In the galaxies and interstellar space far from the stars, where the temperature is not much higher than 3 K and the average density of matter is about 10^7 atoms/m³, large amounts of matter exist in the form of gas and dust particles (Figure 4.3).

The gas and dust form enormous clouds some hundreds of light years across which, in spite of the low density of 10^{-18} kg/m³, have a total mass of 10^{35} kg,

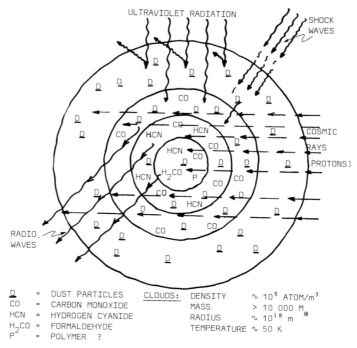

Figure 4.3. The "clouds of life" consist of hydrogen, oxygen, carbon, and dust particles.

that is, 100,000 times the mass of the Sun. There exist very large clouds of up to 10 million solar masses. Such objects move with a velocity of ~100 km/s. The temperature is rather low and does not exceed 100 K.

The earlier discussion about the general course of chemical evolution leads us to suppose that these gas and dust regions contain large numbers of volatile compounds composed of the elements hydrogen, oxygen, carbon, nitrogen, and sulphur (Table 4.1).

The chemical compounds result from the collision of free atoms. But it must be clear that a second collision can reverse the process, disrupting the chemical bond and destroying the molecule. This could occur not only from a collision with another atom or molecule, but also by means of an energetic photon emitted from a neighbouring star. Any photon in the ultraviolet range with an energy greater than 5–10 electron volts can destroy the newborn molecule.

How can the newly synthesised molecules avoid the disruption of ultraviolet radiation? In general there are two possibilities: Either the flux of ultraviolet radiation is low, which would mean the matter is some distance from nearby stars, or the material is shielded in some way. Both "defences" are likely, but the second possibility also introduces a new mechanism which is very favourable for chemical synthesis.

4.2.2 Some of the interstellar molecules exist in solid form

Not all interstellar matter exists in gaseous form. A small part, about one percent, appears in the form of solid dust particles. While some have been observed in the dust-gas clouds discussed above, rather irregular clouds consisting of dust alone have been observed with diameters of 0.01–10 light-years (Fig. 4.3).

The dust clouds are observed in the ultraviolet and visible part of the spectrum with the infrared being much weaker, which indicates that the dust particles are the same size as the infrared wavelength, that is, on the order of 1 μm. The gaseous molecules are only "visible" at radio wavelengths of ~1 cm.

The chemical nature of the dust particles is still not well understood. They seem to consist mostly of silicate with some graphite and metals. Some observers suspect that the dust particles carry a strong positive charge, having lost electrons. There is some evidence that the grains also contain the polymerised formaldehyde, a polysaccharide.

The dust clouds are relatively warm with a temperature of 30 K, cosmic space having a temperature of 2.9 K as the background relict of the "Fireball" (see Chapter 2).

As a result of optical observations it can be concluded that the particles are not of a regular spherical form. It is probable that the process by which dust can act as a shield against ultraviolet radiation, and at the same time as a

Table 4.1. Most Important Interstellar Molecules

		Number of Elements					
		2	3	4	5	6	7
Number of Atoms	13		Cyanodecapentayne ($HC_{11}N$)				
	12						
	11		Cyanooctatetrayne (HC_9N)				
	9		Dimethyl ether (CH_3OCH_3) Ethyl alcohol (C_2H_5OH) Ethyl cyanide (C_2H_5CN) Cyanotriacetylene (HC_7N)				
	8		Methylformate ($HCOOCH_3$) CH_3C_2CN				
	7		Cyanodiacetylene Vinylcyanide acetaldehyde Methylamine				
	6		Methyl sulphide (CH_3SH) Methyl cyanide Methyl alcohol (CH_3OH)	Formamide (NH_2CHO)			
	5	[a] Methane (CH_4) C_4H	Ketene Methanimine (CH_2NH) Formic acid, Cyanoyamine (NH_2CN)				
	4	[a] Ammonia (NH_3) C_3H [a] C_2H	Formaldehyde (H_2CO) Thioformaldehyde (H_2CS)	Isocyanic acid (HNCO) Isothiocyanic acid (HNCS)			
	3	Hydrogen sulfide (H_2S) [a] Water (H_2O) C_2H	Carbonyl sulphide (OCS) Hydrogen isocyanide (HCN) Hydrogen cyanide (HNC) HCO, HNO				
	2	Carbon monosulphide (CS) [a] Carbon monoxide (CO) Nitrogen sulphide (NS) Silicon monosulphide (SiS) Sulphur oxide (SO) Silicon oxide (SiO) Hydroxyl (OH) Nitrogen oxide (NO) Nitrogen sulphide (NS)					

[a] Molecules observed in the atmosphere of Jupiter. *Remark:* Interstellar chlorine and phosphorus compounds have been observed. Total detected: ~60 molecules. The possible number of detectable molecules is limited due to very weak signals: probably less than 150 molecules.

M. Burbidge (1983): "As advances in microwave and infrared techniques reveal the existence of more and more complex organic molecules in dense interstellar clouds, which are likely sites for new star formation, we see a possible primordial "soup" out of which life might have evolved elsewhere than on Earth, in fact anywhere having the right environment".

catalyst for the formation of organic molecules, is of great significance in the study of the chemical evolution of cosmic matter.

It must be stressed that other features of the Universe can strongly influence chemical evolution, for example, comets and meteorites.

4.2.3 Comets: Rare and strange, but formidable, chemical reactors

When comets – relatively small cosmic objects, on the order of 10^{15} kg mass – are far from the Sun they are made up mainly of frozen droplets of water, ammonia, methane, carbone dioxide, and probably hydrogen cyanide, mixed with solid dust particles of graphite, silicon, and iron oxides, and other particles. In other words, their chemical composition corresponds to that generally found in cold cosmic matter.

The origin of comets has not yet been firmly established. In some hypotheses they are very numerous, numbering perhaps 10^{13}; even so, the total mass of these comets is probably less than that of Jupiter. They orbit the Sun, but all within a space not greater than a thousand times the radius of the outermost planet, Pluto. From time to time, due to the gravitational attraction of Saturn or Jupiter, some of the comets are deflected in the direction of the Sun.

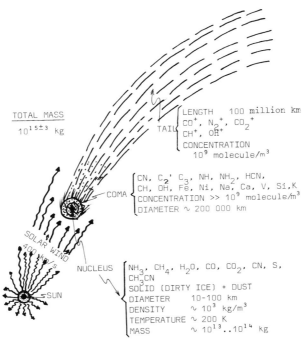

Figure 4.4. Comets have a rather complicated structure (see also Figure 5.1).

When a comet approaches the Sun it has a central solid body, the nucleus, approx. 10 km in diameter, a gaseous and relatively dense coma, plus a rather long but very diffuse tail (Figure 4.4). In the cold, dense state the comet nucleus must be similar to a piece of "dirty ice". It consists of ice mixed with solid methane CH_4, solid ammonia NH_3, solid carbon oxides (CO and CO_2), and dust particles of iron and silicon dioxide. When the comet is within 3 astronomical units from the Sun (1 astronomical unit equals the mean Earth–Sun distance, ~ 150 million km) the nucleus ejects gaseous molecules and radicals such as CN, CH, NH, OH, C_2, C_3 to form the coma. As the comet gets closer to the Sun, the diffuse and very long tail appears, which contains, in addition, the semistable ions CO^+, N_2^+, CH^+, OH^+, H_2O^+. These are produced by the impact of the Sun's ultraviolet radiation and the solar wind.

The chance of collision between comets and the planets, although small, cannot be considered a negligible quantity even for just the Earth. Particularly in the early stages of the Solar System, collision with the Earth could have had an impact on the chemical evolution of the atmosphere and hydrosphere, and maybe on the origin of life. It should be remembered that the mass of the comet is of the order of 10^{15} kg and the mass of the terrestrial atmosphere $\sim 5 \times 10^{18}$ kg. The most recent comet which is believed to have collided with Earth was "Tunguska" in 1908 (see Section 9.4.2).

4.2.4 Meteorites often consist of very "sophisticated" chemical compounds

A less dramatic event is the collision of smaller cosmic bodies consisting not of volatile icy substances but of nonvolatile solid substances. Our Solar System contains a large number of such cold bodies with masses varying from some milligrams to several tons. Large numbers of the smallest approach the Earth every second and are mostly vaporised by their entry into the terrestrial atmosphere at a height of between tens and hundreds of kilometers. When one of these bodies enters the atmosphere, it earns the name "meteor". Meteorites are larger meteors, which may survive atmospheric entry and strike the Earth's surface.

Considering their chemical composition, meteorites fall into two classes:

- iron meteorites (total recorded number ~ 1000), consisting mainly of metallic iron and nickel with some sulphides and carbides;
- stony meteorites (total recorded number ~ 2000), consisting of silicates of magnesium and iron (mostly "chondrites"). It is worth remarking that chondrites include 13 ppG of uranium (ppG = parts per Giga).

Their age has been estimated to be some billion (10^9) years, and they probably originated in some larger, hot cosmic body.

Some meteorites, however, have characteristics which indicate that they have never been heated above 500 K. These meteorites contain not only bonded water molecules but, even more unexpectedly, organic molecules.

4.2 Chemical Synthesis Occurs in Cosmic Space

Some meteorites (e.g., "Allende", 1969) are suspected to be a product of a nearby supernova, occurring prior to the onset of solar system condensation (~ 4.5 billion years ago).

The most extensively studied meteorite of this class is the "Murchison", which fell in the State of Victoria in Australia near the town of Murchison on 28 September, 1969. Within only a few days the first samples were examined in laboratories using advanced techniques. The meteorite contained 2% by weight of carbon and 0.16% by weight of nitrogen. Most impressive, however, is the amino acid content:

Amino acid	Microgram of amino acid per gram of meteorite
glycine	6
glutamic acid	3
alanine	3
valine	2
proline	1
aspartic acid	1
total amino acids	~ 0.4 micromol/gram meteorite

It is worth pointing out that all living matter on this planet, from the simplest virus through the microbes to man, is built up of proteins which are polymers of only some 20 amino acids – the same 20 amino acids for all living things. The Murchison meteorite contained at least six of these 20 amino acids.

The only possible explanation is that these six amino acids are the product of abio-chemosynthesis in an extremely sterile environment without the interaction of any living organism. In other words, the amino acids have been produced by the spontaneous reaction of the most abundant elements, similar to the spontaneous synthesis of complex molecules in the gas-dust clouds such as exist in the centre of our Galaxy.

4.2.5 "Organic molecules" on the Moon and planets

We belong to the generation of the Moon explorers. One of the objectives of the scientific investigation of our satellite was to determine the chemical and geological properties of our nearest cosmic neighbour.

In fact, the Moon's crust is similar to that of the Earth. The most important difference is the almost complete absence of organic substances so characteristic of our planet. However, it is not true to say that there are no organic substances on the Moon.

The very complex investigation techniques, and the no less complex discussions on the interpretation of results, seem to point towards the following: The surface of the Moon contains a very few organic substances, among them some amino acids in the proportions listed below.

Amino acid	Nanograms of amino acid per gram of Moon rock
glycine	~10
alanine	~1
glutamic acid	–
valine	–
proline	–
aspartic acid	~1
serine	~1
leucine and isoleucine	~1
lysine	~1

These amounts are approximately one thousand times smaller than the concentrations of amino acids in the Murchison meteorite.

We should also investigate the presence of organic substances on other planets. Here our knowledge has increased dramatically in the last decade and should increase still further in the coming decades. Nevertheless, the level of information is still very low. A simplified review of the chemical properties of the planets is given in Table 4.2.

Table 4.2. Composition of the Planets

Planet mass of Earth =1.0 ϱ=density	Mean temperature/relative solar radiation: I.P.[a]	Spheres			
		Solid	Liquid	Gaseous	Biotic
Mercury 0.054 ϱ=5.3	550 °C/6.6 (day side)	Silicates of Al, Fe, Al, Ca (Pyroxene)	No	No	No
Venus 0.814 ϱ=4.95	~470 °C/1.8 (day side) I.P.?	Silicates of Al, Fe, Mg, Ca (metallic core) Volcanic activity?	No	Pressure: 100 bar; 99% CO_2 1% N_2, SO_2 0.4% H_2, H_2SO_4, HCl	No
Earth 1.000 ϱ=5.52	282K $\dfrac{\sim 15\,°C}{1.0}$ I.P.=0.04	Silicates of Al, Fe, Mg, Ca (Olivine, Pyroxene) (Fe, Ni core)	Water: 2/3 of the surface	Pressure: 1 bar; 78% N_2, 21% O_2, 0.03% CO_2	Yes

4.2 Chemical Synthesis Occurs in Cosmic Space

Table 4.2. (Continued)

Planet mass of Earth =1.0 ϱ=density	Mean temperature/relative solar radiation: I.P.[a]	Spheres			
		Solid	Liquid	Gaseous	Biotic
Moon 0.107 ϱ=3.34	$\dfrac{143\,K}{1.0}$	Silicates	No	No	No
Mars 0.107 ϱ=3.95	$\dfrac{-260\,K}{0.43}$ I.P.?	Silicates, Olivine + 50% FeO	No	Pressure: 0.01 bar 99% CO_2, 1% N_2, CO_2, H_2O	No No
Asteroids 0.001 (Total mass)	~200 K/0.1	Solid rocky	No	No	No
Jupiter 317.45 ϱ=1.33	127±3 K/0.037 I.P.=400	Only part of total mass (14 Earth's masses) Silicates, Hydrogen	Probably liquid Ammonia (NH_3) and Methane (CH_4)	25% H_2, 15% He NH_3, CH_4, H_2O (also: SiH_4, PH_3, H_2S, CH_3NH_2, C_2H_6, HCN, GeH_4, C_2H_4)	No
Saturn 95.06 ϱ=0.68	97±3 K/0.011 I.P.=200	Only small part of total mass (~16 Earth masses)	No	88% H_2, 11% He, 0.1% H_2O, 0.1% CH_4 + NH_3, PH_3, C_2H_6	No
Titan, Saturn's Moon ϱ=1.9	~90 K	Solid: 52% Rock, 48% Ice	Methane rain, Methane rivers	Pressure: 1.5 bar, N_2, Argon, H_2 CH_4 C_2H_6, C_2H_4, C_2H_2, HCN, HC_3N	No
Uranus 14.50 ϱ=1.56	58±3 K I.P.=0	Only part of total mass: Silicates, Ice, Methane	No	H_2, CH_4, He	No
Neptun 17.60 ϱ=2.27	53±3 K I.P.=3	Only part of total mass: Silicates, Ice, Methane	No	H_2, CH_4, C_2H_6, He	No
Pluto 0.002 ϱ=0.7	<40 K	Solid "gases" ϱ<1500 kg/m^3 Methane, Ammonia	No	No	No

[a] I.P. = internal power (in PW = 10^{15} watt).
ϱ = density, kg/litre.

4.3 The Origin of the Planets

4.3.1 Have the planets been formed "by chance"?

We are creatures of this planet, and yet it is surprising how little we know about its origin and that of the neighbouring planets. There are at least three explanations for this:

a) Up to now we know of only one planetary system – our own. Neither direct observation with the most advanced astronomical methods nor indirect deductions give any information on the existence of other planetary systems. Equally, our own system could not be detected by an observer close to the nearest stars some 5–10 light-years away, using the methods available to us. Perhaps in the coming decades extraterrestrial observatories will discover other planetary systems. The richness of the Universe as it is known to us, however, makes highly unlikely the suggestion that our planetary system is unique.

The number of large planets we do know, their satellites, the presence of hundreds of planetoids, probably hundreds of thousands of larger meteorites and millions of comets, contradict the idea that our system is exceptional and all speak for the high probability of a general mechanism by which planetary systems may be formed.

b) Only a few members of our planetary system are accessible for direct and detailed examination, such as a full chemical and mineralogical analysis, and continuous measurement of the most important parameters such as temperature, atmospheric pressure, heat flux, and magnetic field.

c) Even these planets have been investigated at their surfaces only. The Earth itself has been examined directly to only about about 12 km below the surface. All our knowledge about the deeper layers is inferred from indirect measurements based, for example, on the propagation of earthquake waves.

Having noted the recent dramatic jump in our knowledge coupled with the realisation that the depth of that knowledge is, however, not great, we can also draw the conclusion that knowledge of general planetary evolution will lead to a fuller understanding of our own planet's birth, and that, in turn, may be crucial for the further development of mankind on this planet. The present hypotheses of the planets' evolution cannot even answer the following questions:

– Are planets formed in a catastrophic event such as the explosion of a supernova in the neighbourhood of a young double-star system (a "big bang" theory of the origin of the Solar System)?
– Are planets formed as a "normal" product of the evolution of the proto-solar nebula or from gas-dust clouds?

- Do planets originate from a certain class of stars and only during a certain period of their evolution? How numerous are these stars and what is the probability that a planetary system will be formed?

4.3.2 The protoplanet, the first stage of evolution

In spite of the lack of firm theories, it is generally accepted that all the planets, among them the Earth, at the time of their birth approx. 4.5×10^9 years ago, were in a cold rather than a hot state and in small rather than large fragments.

There is a lot of evidence for this. One basis is the fact that the terrestrial abundance of the nonvolatile elements such as silicon, magnesium, aluminium, calcium, and iron is much higher than the equivalent cosmic abundance. Similarly, the terrestrial abundance of the volatile elements helium and neon, in particular, but also hydrogen, carbon, and nitrogen, is many orders of magnitude lower than the cosmic abundance. For example, helium is 10^{14} times and krypton 10^7 times less abundant on Earth than in average cosmic matter. Other volatiles, of course, do not show such extreme differences, but in all cases the terrestrial abundances are lower.

These observations can be explained as follows. In the first stage of a planet's evolution, the matter in the protosolar nebula would be in the form of small, cold, solid dust particles consisting of the nonvolatile chemical compounds such as oxides of silicon, magnesium, aluminium, iron, and also oxide of hydrogen (water) in the form of ice. Other solidified hydrogen compounds, such as methane (CH_4) and ammonia (NH_3), would also have been present. The temperature of these dust particles was approximately 100 K. The noble gases helium, neon, krypton, etc. remained in gaseous form surrounding the dust particles. All these noncondensable gases have been dissipated into cosmic space.

The accretion of the small, cold dust particles to form large solid bodies is due to gravitational attraction, and in packing together, the dust particles release their gravitational energy in the form of heat. Roughly calculated, the temperature in the centre of the proto-Earth was probably 1750 K, which is enough to decompose some of the chemical compounds, releasing ammonia and water, and later leads to melting of some of the material. The molten compounds include the simple silicates of aluminium, magnesium, and metallic iron.

This molten metallic iron, together with iron sulphide, being denser than the molten silicates, moves towards the Earth's centre. This is the first main chemical separation – the so-called "iron catastrophe".

The first chemical differentiation, that of the heavier iron components and the lighter silicates, which includes most of these radioactive nuclides, had an important impact on chemical evolution on this planet.

4.3.3 The chemical evolution of the Earth: A complex and dramatic development

Even during the early stages of the planet's evolution, chemical processes play an important role. Some lead to separation of the chemical elements and compounds, others to mixing. It can be said that in the past the processes of separation were more effective than the processes of mixing. The Earth is now, and has been for about 4 billion years, a very heterogeneous body with very large chemical differentiation. The separation can be considered to result in three spheres:

- the gaseous sphere – the atmosphere – with only a few characteristic components;
- the liquid sphere – the hydrosphere – containing principally water, the most abundant chemical compound in the entire Universe;
- the solid sphere – the lithosphere – containing without exception all the stable and semi-stable nuclides which exist in Nature.

It is worth making some comment on the history of these spheres beginning with the most massive, the lithosphere (Figure 4.5). Here the term "lithosphere" corresponds to the total sphere of the Earth's solid material.

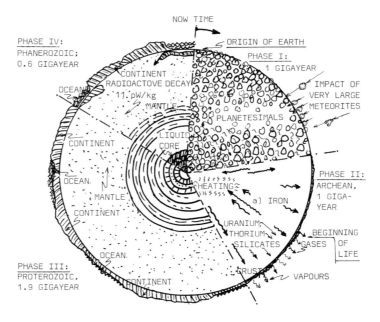

a) "IRON CATASTROPHE": RELEASE OF 2×10^{30} J THIS IS THE SOURCE OF THE MAGNETIC FIELD OF EARTH.

Figure 4.5. Four stages in the evolution of the Earth.

4.3 The Origin of the Planets

The first stages, up to the iron catastrophe, covering some tens of millions of years, have already been briefly discussed (see Figure 4.5, Phase II).

The differentiation of the lithosphere continues, and results in the formation of three layers:

1) the crust;
2) the mantle, sometimes further subdivided into the upper and lower mantle;
3) the core, divided into the outer and inner core (Figure 4.6).

In parallel with the sinking of the metallic iron, the lighter components flow upwards, the separation processes increasing with chemical and mineralogical differentiation. Part of the residue of the magma crystallization in the mantle forms the so-called hyperthermal solution, in which some of the less abundant elements concentrate. Typical examples are lithium, beryllium, boron, fluorine, and tantalum. It is worth noting that some of these light elements still have a role to play today and in the future as components in the development of fusion energy.

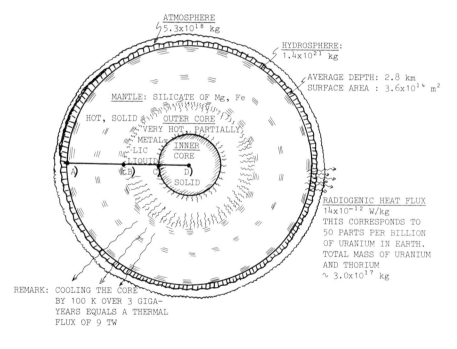

Figure 4.6. The structure of the Earth is highly differentiated.

The upper layers of the lithosphere are strongly influenced by the action of the atmosphere and hydrosphere. One prominent example is the dissolution of sodium in the form of sodium chloride and its extraction from the rocks and transfer to the oceans. Other alkali metals, such as potassium, are absorbed by the newly formed minerals and are chemically bonded in a nonsoluble form. The extraction processes can be so efficient that the elements become enriched in the nonsoluble form, forming, for example, aluminium oxide in bauxite. These processes are even more extensive when the biosphere is considered.

Many of the water-soluble substances, such as chlorides, sulphates, carbonates, and phosphates, are later precipitated and form an "inactive store".

A further example of the intensity of the separation occurs when even isotopes of the same element are separated. A well-known example is the oxygen isotopes oxygen-18 and oxygen-16. In some sediments the ratio of $^{18}O/^{16}O$ is larger than in water. However, we must not underestimate the role of the mixing processes. Here again, isotopic variations can be used as an example.

We know, for instance, that some isotopes of lead were not entirely produced during the period of element synthesis, but resulted from the radioactive decay of thorium-232, uranium-235, uranium-238, and even short-lived neptunium-237, over a long period of time. Nevertheless, most minerals of lead show the same isotopic composition, which can only be explained by an effective mixing of all the various isotopes, regardless of origin.

The rock masses that form the external part of the Earth have been in rapid motion for hundreds of millions of years. This motion is so rapid that 2/3 of the crust has been recycled into the Earth's mantle in the last 200 million years.

It is important to see that some very significant processes in the lithosphere result from the heat released by the radioactive elements, mostly thorium and uranium. The following geological processes have been driven by the radiogenic heat:

- the volcanic activity: the water of the oceans and the nitrogen and carbon dioxide of the primitive atmosphere are products of volcanic activity;
- the "iron catastrophe" (see Figure 4.5): the terrestrial magnetic field is driven by the "iron catastrophe";
- the drift of continents, influencing the global climate, is also an effect of radiogenic heat (see Chapter 6).

We see, therefore, that chemical processes in the lithosphere are moving in both directions, increasing heterogeneity, by separation mechanisms, and, parallel to this, increasing homogeneity by mixing effects. Important questions must, however, remain unanswered. At the present time, what is the relative strength of these two opposed processes? Which is dominant, the

4.3 The Origin of the Planets

mixing or the separation process? How has human activity influenced them? Can mankind control both processes and so overcome the risk of a "resource catastrophe"?

Before discussing the evolution of the hydro- and atmosphere, here are some additional remarks concerning the chemical composition of the Earth's crust.

4.3.4 All stable elements present in the Universe exist on Earth

What is the chemical composition of the Earth today? A simple question, and a simple answer, might be expected. This is not the case. Direct chemical and mineralogical analyses are only possible on samples which can be brought to the laboratory, and drilling on the continents has not exceeded a few kilometres depth. The thickness of the continental crust is, however, about 20 km. The same is true of the oceanic crust. The centre of the Earth is still a

Table 4.3. Chemical Composition of the Earth's Crust (Content of 1 Mg of crustal rocks $\sim 0.357\,\text{m}^3$)

	Element		Kilograms		Element		Grams
1	Oxygen	$_8$O	446.	22	Rubidium	$_{37}$Rb	90
2	Silicon	$_{16}$Si	227.	23	Nickel	$_{28}$Ni	72
3	Aluminium	$_{13}$Al	81.	24	Zinc	$_{38}$Zn	80
4	Iron	$_{26}$Fe	58.	25	Cerium	$_{58}$Ce	60
5	Calcium	$_{20}$Ca	36.	26	Copper	$_{29}$Cu	55
6	Sodium	$_{11}$Na	28.3	27	Yttrium	$_{39}$Y	35
7	Magnesium	$_{12}$Mg	27.7	28	Lanthanum	$_{57}$La	30
8	Potassium	$_{19}$K	16.8	29	Neodymium	$_{60}$Nd	28
9	Titanium	$_{22}$Ti	8.6	30	Cobalt	$_{27}$Co	28
10	Hydrogen	$_1$H	1.4	31	Scandium	$_{21}$Sc	22
11	Phosphorus	$_{15}$P	1.1	32	Nitrogen	$_7$N	20
12	Manganese	$_{25}$Mn	0.95	33	Niobium	$_{41}$Nb	20
13	Fluorine	$_9$F	0.465	34	Lithium	$_3$Li	20
14	Barium	$_{56}$Ba	0.425	35	Gallium	$_{31}$Ga	15
15	Strontium	$_{38}$Sr	0.375	36	Lead	$_{82}$Pb	10
16	Sulphur	$_{16}$S	0.300	37	Boron	$_5$B	10
17	Carbon	$_6$C	0.200	38	Thorium	$_{90}$Th	10—13
18	Chlorine	$_{17}$Cl	0.190	39	Tin	$_{50}$Sn	1.5
19	Zirconium	$_{40}$Zr	0.165	40	Uranium	$_{92}$U	2.4—4
20	Vanadium	$_{23}$V	0.135	41–87	other Elements		≪ 2
21	Chromium	$_{24}$Cr	0.096				
							Milligrams
				88	Mercury	$_{80}$Hg	20
				89	Platinum	$_{78}$Pt	5
				90	Gold	$_{79}$Au	2

long way away, more than 6300 km. Indirect methods such as seismology do, however, permit geophysicists to build up a picture of the chemical and mineralogical composition of the whole Earth. The following are some of the most important features.

- The Earth contains, without exception, all of the stable and semi-stable elements; also all stable and semi-stable isotopes are present.
- The crust has the richest chemical composition.
- The most important component of the Earth's crust is oxygen, particularly oxygen-16, being a double-magic nuclide with 8 protons and 8 neutrons. About 46.6 percent of the crust by weight is oxygen. This corresponds to 62.5 atom-percent; that is, of every 1000 atoms in the crust 625 are oxygen. Because oxygen forms anions, O^{2-}, with a relatively large volume, the proportion of oxygen by volume is 92%. The Earth's crust is practically an "oxygen sphere" in the form of solid oxygen compounds.
- The next most frequent component of the crust is silicon, $\sim 27.7\%$ by weight. Most of this element exists in the form of silicates with the general formula $(a \cdot \text{MeO}) \cdot (b \cdot \text{SiO}_2)$, where Me represent any metal, and a, b are variables.
- Further down the scale lie the six metals: aluminium, iron, calcium, sodium, potassium, and magnesium. All the elements above, including the six metals, represent 98.5% of the crust by weight.
- The remainder, that is, 1.49 weight-percent, is taken up by 82 elements ranging from 0.10 weight-percent for hydrogen, phosphorus, and manganese, up to thorium and uranium, with 15 parts per million (that is, 0.0015%), and mercury and platinum with 15 parts per billion (see Table 4.3).

We have covered the Earth's crust here, but it should not be forgotten that the crust contains only about 4×10^{22} kg of material, that is, only 1 weight-percent of the Earth's total mass.

We know, or rather we can deduce, a few things about the chemical composition of the interior of our planet. The most accepted hypothesis considers the core of the Earth to be iron, which, in turn, gives the chemical composition of the Earth as a whole:

iron	34.6 weight-percent
oxygen	29.5 weight-percent
silicon	15.2 weight-percent
magnesium	12.7 weight-percent
nickel	2.4 weight-percent
all other elements (~ 87)	5.6 weight-percent

The core contains iron and nickel, but probably also some light elements such as oxygen (~ 10 weight-percent) or sulphur (9–12 weight-percent), which

4.3 The Origin of the Planets   77

are "dissolved" in the liquid iron core. The presence of these elements in the core allows the core to be in chemical equilibrium with the overlying mantle and the Earth as a whole to reflect the distribution of the heavy elements in the Sun.

4.3.5 The history of the Earth has been influenced by the movement of the continents

The solid ground under our feet is not as stable as one might think. On the contrary, we have come to realise that the continents are moving and have done so for at least 500 million years. Similarly, a good part of the ocean floor seems to have been formed only within the last hundred million years.

Approximately two-thirds of the solid surface of the Earth is relatively young. The older ocean floor seems to have been subducted and destroyed in the deeper layers of the Earth's crust.

COMPONENTS	UNIT	CRUST	
		CONTINENTAL	OCEANIC
PERCENT OF SURFACE	%	40 - 45 %	55 - 60 %
HYDROSPHERE DEPTH	km	---	4.3
SEDIMENTS DEPTH	km	2	2
DENSITY	$\frac{1000 \text{ kg}}{m^3}$	2.45	2.45
CRUST THICKNESS	km	10 to 70	MEAN 6
ROCK		GRANITES	BASALTS
DENSITY	$\frac{1000 \text{ kg}}{m^3}$	~2.8	~3.
HEAT FLOW	$\frac{mW}{m^2}$	~60	~120
HEAT PRODUCTION	$\frac{W}{kg}$	0.023	0.0063
AGE	YEARS	~ 3.8×10^9	< 2×10^8
TEMPERATURE AT 50 km	°C	450 - 800	600 - 1200

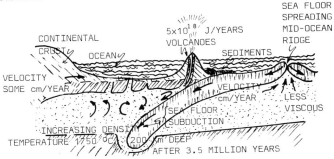

Figure 4.7. The continental and oceanic crusts have rather different properties.

The continental plates move with a velocity of some centimetres per year in different directions, e.g., America moves away from Europe and Africa. This is due to an up-flow of basaltic magma which forms the new sea floor (Figure 4.7; see also Figure 6.9).

The old sea floor has been buried in the deeper and hotter layers of the outer mantle down to 300 km or more. In these hot layers, with temperatures higher than 1700 °C, the old rocks melt and dissolve. At the present time six large continental plates and some other smaller plates take part in this global movement.

This model can explain a number of local phenomena, for example: earthquake centres, the nature and origin of the sea floor, the origin of mountains and volcanoes, and magnetic anomalies.

All this movement can only occur as a result of the input of energy, most likely in the form of a high-temperature heat flux. This transfer of heat to the mechanical energy moving the continents is a beautiful example of the heat engine which makes up the Earth's crust.

Allowing for a few unanswered questions, the most obvious source of this heat flux is the heat of radioactive decay of the semi-stable, long-lived nuclides thorium-232, uranium-235, uranium-238, and potassium-40. Earlier, other radioactive nuclides would also have played a significant role.

At the present time the heat flux due to radioactive decay is only 57 TW (57×10^{12} W) with an additional approx. 9 TW coming from primordial heat. Together this equals ~ 66 TW, that is, only 0.04% of the flux of energy arriving from the Sun (see Chapter 6). In addition, only part of this radioactive heat flux can be transformed into mechanical energy, probably only a few percent. The energy of earthquakes averages about 0.3 TW, that is, approx. $\frac{1}{2}$% of the total geothermal flux.

Looking back, we can now see how the products of supernova explosions five billion years ago, thorium and uranium, have influenced the movement of continents and hence the evolution of the climate which, in turn, has affected the whole development of mankind up to now, and will continue to do so in the future.

4.3.6 The first phases of chemical evolution were driven by different energy sources and were influenced by a number of factors

The most important evolutionary steps are the chemical processes. The chemical reactions require energy supplies of the order of electron volts per molecule, that is, $\sim 10^{-19}$ J/molecule. The first question is, what energy sources have provided the necessary driving force? The second question is, at what temperature would the different reactions have occurred? The temperature governs the speed of reaction. In the first period of the Earth's formation, for example, the surface temperature reached 300–400 K. This temperature is rather low so the reactions proceeded rather slowly, therefore

4.3 The Origin of the Planets

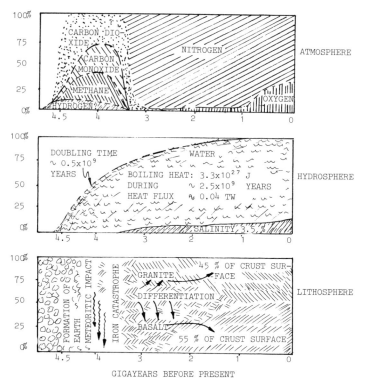

Figure 4.8. Evolution of the three spheres.

their effectiveness was small. That the reactions, in fact, proceeded more quickly can be explained by the presence of chemically active substances or catalysts. The third question, then, is, what kind of simple catalytic substances could have been present during the early stages of terrestrial chemical evolution? (Figure 4.8)

Going back now to the question of the source of free energy, the simple but incomplete answer is the Sun. The flux of solar energy reaching the Earth today is more than 175 PW (PW = petawatt = 10^{15} W). What flux of solar energy should we assume for the period 5×10^9 years ago? We need to know not only the magnitude of the flux of solar radiation, but also the spectral distribution of this radiation, that is, the proportion of ultraviolet, visible, and infrared radiation. This information alone is still not sufficient for a complete characterisation of the properties of the solar flux, since it is also strongly dependent on numerous properties of the terrestrial atmosphere, including pressure, chemical composition – especially the concentration of water vapour and carbon dioxide and dust particles. More or less reasonable assumptions give a resulting temperature on the Earth's surface in this first period of between 0 °C and 150 °C.

The problem of the solar radiation spectrum is very difficult. At the present time, the ultraviolet radiation equals approx. one-hundredth of the total flux, or 3×10^{18} photons/m² · sec, since the ultraviolet radiation is absorbed by the ozone layer in the atmosphere (see Section 5.2.3). There is a strong opinion that the early terrestrial atmosphere, however, did not contain free oxygen, therefore the ozone layer was also absent, which means that the ultraviolet radiation reached the Earth's surface. In such circumstances the ultraviolet radiation would have been the strongest and most chemically active source of energy, but not, however, the only energy source on the primitive Earth. The second energy source which probably played a role was terrestrial radioactivity (see Section 6.4.1). The present flux of energy from radioactive decay is 100 mW/m² (mW = milliwatt). In the distant past it would have been higher, say as much as 0.5 W/m². This is still very low relative to the total solar energy of 200 W/m² or even the ultraviolet part, 2 W/m². Nevertheless, this factor-of-ten difference does not diminish its role, especially in some particular local regions of radioactive influence.

Some part of the decay heat appears at the Earth's crust in the form of volcanic activity. These local sources of heat, with temperatures as high as 1000 °C in the flowing lava, should not be underestimated as a factor in the

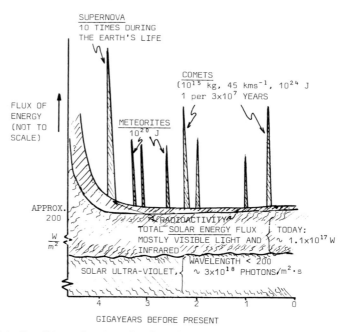

Figure 4.9. Candidates for the role of the intensive and/or extensive sources of free energy in the past. (*Remark*: In the past there was a significantly smaller energy flux resulting from the evolution of the Sun as a Main Sequence star.)

4.3 The Origin of the Planets

chemical evolution. There are, however, still some other candidates for sources of free energy.

One is the meteorites (see Section 4.2.5). A large meteorite with a mass of a 10^{12} kg falling onto the Earth can achieve a kinetic energy of approx. 10^{20} J, equivalent to the total solar flux during one hour, and this enormous amount of energy is concentrated in a rather small space. The local impact of such an explosion caused by a falling meteorite is clearly of great significance, and the chemical changes in the surrounding matter – solid, liquid, and gaseous – would have been dramatic (Figure 4.9).

Another cosmic source of energy on this planet is comets (Section 4.2.4). The consequences of a collision with a comet can be of greater significance than one with a meteorite, not only because the mass and entrance velocity are higher than for meteorites, with the final kinetic energy reaching approx. 10^{24} J, but also because of the internal chemical make-up of the comet itself. As we have seen, the comets in the neighbourhood of the Sun contain a large amount of chemically reactive substances. It is clear that an impacting comet can be a very formidable chemical "factory", producing a large variety of reaction products in a very short span of time and in a very concentrated form over a limited area of the Earth's surface.

The probability of a collision with a comet having a mass of 10^{15} kg is small, about one event every 10 million years. Is this frequent enough to influence the chemical development of our planet? How many such collisions have, in fact, occurred?

There are some hypotheses which suggest that, due to the motion of the Solar System around the centre of the Galaxy, with approximately one revolution every 250 million years, it passes several times through the spiral galactic arms which are rich in dust particles and perhaps also in larger cold bodies which would form comets. If this is what really happens, then comets could play an even more significant role. But how sound are these ideas? Do they have any basis in reality or are they merely an academic exercise?

In the search for sources of energy and matter for the terrestrial chemical processes we must include the most spectacular of all galactic events, the explosion of a star, or supernova.

A typical galaxy averages one supernova every 50 years. The total energy released is approx. 10^{43} J within less than one year's time span. This corresponds to a power of 10^{35} W. If we remember that a typical star such as the Sun has a power of 10^{26} W, then a supernova emits energy with a power equal to that of a whole galaxy containing billions of stars.

4.4 Synthesis of Complex Molecules on the Primitive Earth

4.4.1 The primitive atmosphere includes mostly molecules containing hydrogen

Before going more deeply into the mechanisms of chemical evolution, we present a short resume of the initial conditions.

The primitive Earth had a chemical composition rather different from the mean composition of the protosolar cloud. The most characteristic feature was the unexpectedly low level of hydrogen in the free, unbonded form. The relatively weak gravitational field of the Earth was not able to prevent this hydrogen from escaping into space. In a rather short period, on the order of 1000 years, most of this hydrogen was lost. The small amounts which remain result from continuous emission from volcanoes and degassing from the Earth's crust.

Small amounts of ammonia originate from the same sources. The amount of ammonia in the atmosphere is limited not by dissipation from the Earth's gravitational field but by continuous destruction by the ultraviolet component of solar radiation. Ammonia is a rather weakly bonded molecule, and the ultraviolet radiation splits it up into free nitrogen (N_2) and hydrogen (H_2). The reversal of this process is practically impossible, requiring the existence of high pressure and a source of free energy. In brief, ammonia has a very short lifetime in the Earth's environment.

Active volcanoes, resulting from the heating of the Earth's crust due to the radioactive decay of uranium, thorium, and other unstable elements, are a source of other volatile compounds, mostly hydrogen-containing molecules. Including those already discussed they include:

- hydrogen H_2
- ammonia NH_3
- water H_2O
- methane CH_4

- hydrogen sulphide H_2S
- hydrogen cyanide HCN
- formaldehyde H_2CO

plus the hydrogen-free compounds:

- nitrogen N_2
- carbon oxides CO, CO_2
- cyanogen N_2C_2

Most of these volatile compounds are heavy enough to be retained by the gravitational field of the Earth and stable enough not to be decomposed too quickly by ultraviolet radiation. For this reason, with the exception of hydrogen, helium, and ammonia, these compounds accummulated with time and formed the primitive gaseous atmosphere and, somewhat later, the liquid hydrosphere (see Figure 4.8).

These are the basic theories about the composition of the primitive Earth and the role of the different sources of free energy such as solar radiation, radioactivity, the impact of objects, and other cosmic events.

4.4.2 The amino acids; their ease of synthesis

Take a mixture of some hydrogen-containing molecules, such as water, methane, ammonia, and hydrogen sulphide, over a rather wide concentration range and expose it to a source of free energy in the form of electromagnetic radiation, for example, ultraviolet, heat, radioactive energy (alpha- or beta-particles or gamma-photons), or shock waves, and the result is rather surprising.

Significant amounts of different amino acids can be synthesised, including not only simple but also more complex molecules. Most of these amino acids were earlier noted to exist in cosmic objects and in terrestrial living organisms. These amino acids, which form the cornerstones of terrestrial life and have been found in meteorites, are thought to have been synthesised in the primitive atmosphere. On this last point, most of them have in fact also been synthesised in laboratory experiments in which the conditions of the primitive atmosphere were simulated. In the primordial atmosphere similar, spontaneous syntheses undoubtedly occurred. The main questions, however, are: What was the "productivity" of the various processes and what was the subsequent history of the newly synthesised amino acids? This productivity of abiogenic syntheses depends very strongly on the postulated chemical composition of the primordial atmosphere and on the type and intensity of the flux of free energy.

Some calculations give a production rate as high as $5 \, mg/m^2 \cdot day$, which corresponds to $2 \, g/m^2 \cdot year$, or $2000 \, kg/m^2$ in a million years. Even allowing for the fact that parallel decomposition of the amino acids has not been included, this result appears to be much too high.

It must also be said that a value as high as $5 \, mg/m^2 \cdot day$ does not correspond to a particulary high efficiency of transformation, which can be seen from the fact that the mass of the atmosphere on one square meter of the Earth's surface is approx. 10,000–20,000 kg. Related to a daily production of $5 \, mg/m^2$ of amino acids, this represents the transformation of less than one part per billion (1 in 10^9). The laboratory experiments achieved a similar order of magnitude. In part, the low efficiency is due to the inefficiency of the available sources of free energy in promoting these reactions.

We must remember that the effectiveness of spontaneous natural reactions should not be measured by human scales. Nature has a lot of matter, a lot of free energy, and, most important, a lot of time. Even with a very low efficiency a large quantity of material and very complex products can eventually be produced. All bases found in the nucleic acids of living creatures on Earth can be produced in a primitive atmosphere of methane, nitrogen and water by impact of electrical discharges.

4.4.3 How were the large molecules, the polymers, produced?

It seems to have been established that the small molecules forming the component parts of living matter were produced by spontaneous chemical synthesis. Is this the simple solution to the mystery of the origin of life on our planet?

A living being on Earth has an extremely complex structure in which the components are not the rather small molecules such as sugars, amino acids, and nucleotides, but very large molecules. The polymers are made up of thousands or even hundreds of thousands of smaller molecules, the monomers. Below, this relationship is shown diagramatically.

From this simple scheme we can also see that, in conditions where large numbers of water molecules are present, the polymers can also be broken down again into the smaller molecules (called depolymerisation). This raises the problem of how, in fact, the prebiotic, spontaneously synthesised monomers (nucleotides, amino acids, polysugars, fatty acids) could have been polymerised. If this occurred, as seems likely, in aqueous media, then not the polymerisation process but only the depolymerisation occurs spontaneously. This stage is still a weak link in the understanding of the whole evolutionary chain and requires further investigation.

Here is the relationship between these components:

				Examples of polymers	Appropriate monomers
Living organism, e.g., unicellular		Structures, e.g., nucleus	Polymers	polynucleotides (nucleic acids)	nucleotides, e.g., adenosine; phosphate: thymidine-phosphate
		Structures, e.g., membranes, organelles	Polymers	polypeptides (proteins)	amino acids: glycine, alanine, together $\sim$ 20 amino acids
				polysugars (cellulose, starch)	hexose, pentose
	Intracellular aqueous media		Water	lipids	fatty acids, e.g., palmitic acid, oleic acid
			Monomers		
			Simple molecules and ions:	NaCl	
				Na^+, Cl^-	

The process of polymerisation proceeds as follows:

$$\begin{pmatrix} \text{monomer} \\ \text{(large number} \\ \text{of monomers)} \end{pmatrix} \underset{\text{depolymerisation}}{\overset{\text{polymerisation}}{\rightleftarrows}} \begin{pmatrix} \text{polymer} \\ \text{(one macro-} \\ \text{molecule)} \end{pmatrix} + \begin{pmatrix} \text{large num-} \\ \text{ber of water} \\ \text{molecules} \end{pmatrix}$$

It is possible for polymers to be synthesised directly, bypassing the steps of monomer link-up, but this too has been insufficiently studied.

Yet even if these difficulties could be satisfactorily overcome, the existence of polymers does not mean that life has begun. The path from polymer synthesis to the beginning of life must be very complex and difficult, and each stage will have a very low probability. In spite of these limitations life did evolve from nonliving matter, starting off a whole new phase of evolution. But what is life?

4.5 What Is Life? The Need for a General Definition

4.5.1 Could life have originated spontaneously?

Of one thing we can be certain — at least one system of life exists in the Universe. There are almost limitless possibilities for coupling this one fact with different opinions, hypotheses, or faiths; for example, the following combinations:

- terrestrial life is unique in the Universe, or else life is present in other parts of the Universe, other parts of the Galaxy, or even other parts of the Solar System,
- life originated spontaneously on this planet without interaction with any other extraterrestrial living system, or is a product of extraterrestrial, live, or supernatural forces.

These ideas have been disputed for a very long time and it is not expected that any solutions will be found in the near future. Nevertheless, the case is presented below for life originating spontaneously as a sequence of steps in the chemical evolution of cosmic matter which culminated in the appearance of life some 5 billion years ago. These arguments are also influenced by the fact that all terrestrial living organisms, from the virus up to man, are built according to the same basic plan.

Can the same be true of other living organisms we might meet one day in other parts of the Universe? What, indeed, is life, from the cosmic rather than just the narrow terrestrial point of view?

4.5.2 The physical aspect of life

The language of physics claims to be a general, not just a specifically terrestrial language. The laws of nature formulated by physics are assumed to be valid not only here and now but also elsewhere and at all times, therefore we should use this language for a description of life in a neutral way, avoiding the peculiar terrestrial properties of life, which is, however, the only known example on which we can draw.

The first property of life seems to be spontaneity, the ability to evolve from a nonliving system. According to the well-established Copernican principle that our planet is not a unique, special, or central point in the Universe, we must assume that life is also present not only here, but elsewhere.

Life, according to these ideas, is therefore characterised by spontaneity and universality. But what is life?

The next property is order. The words organism, organisation, and order are not only semantically but also physically coupled. The living being has a very highly ordered state. Disorder leads to decay and death.

Our knowledge of the physical world, however, shows that the ordered state spontaneously transforms to the disordered state. The flow of events (the passage of time) is directly connected with the increase of disorder and the decrease of order. This is the foundation of the second law of thermodynamics. There is only one way to counteract this destruction of ordered systems, and that is to add an amount of "ordered" energy which reconstitues the ordered structure. Of course, the energy itself is derived from the transformation of order into disorder elsewhere in the system.

From this we derive a similar requirement for the living system which, as an example of a highly ordered structure, experiences the destructive influence of the passage of time. The only way for order to be conserved is to add to the living system an appropriate portion of ordered energy in an appropriate form. Of course, this requires that the source of energy is available for as long as life exists. The need for an ordered input of energy is therefore the third property of the living system.

We have seen here that the continuous maintenance of internal order in the living system requires not only a portion of ordered energy but also an appropriate flux of energy. The intensity of the energy input must be neither too small nor too large. Each ordered level needs a matching flux of energy. The same is true of the energy which is released from the system. The intensity matches the degree of order in the system where it is generated. The regulation of these energy process and material flows requires a very complex control system, and the existence of such a control system for energy and material transformations is the fourth property of any living system.

The environment of the living system changes with time: the source of the energy input, the energy flux, the material medium of the living system, all these parameters change both in the short term and over longer periods of

4.5 What is Life? The Need for a General Definition

cosmic time. Life can be conserved only if it can adapt to the changing environment. Life has to evolve. The changes to the living system are possible when the less well adapted organisms are eliminated and the better adapted creatures multiply. The fifth and last property of a living system is the ability to multiply, for self-reproduction, not only in an identical form but also with some slight changes, some new, detailed adjustments.

Summarising, we can consider in these very general terms that life here and in other parts of the Universe must have the following basic properties:

1. spontaneous creation from nonliving matter;
2. a very highly ordered structure;
3. a constant flux of ordered energy from external sources;
4. a very effective internal control and regulation system which responds to external situations;
5. the ability to multiply (self-reproduction) (Figure 4.10).

It must be stressed that up to now, in attempting to describe the properties of living matter, we have not used any information concerning the kind of matter from which living systems are built or the kind of ordered energy which has to be consumed. Up to now the discussion has been kept very general.

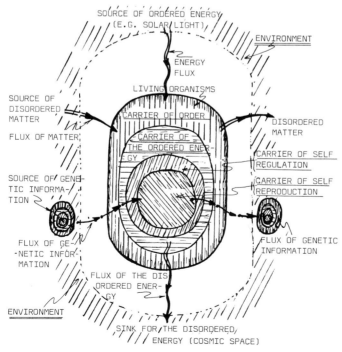

Figure 4.10. Living organisms described very schematically.

4.5.3 What kind of elementary forces can play the role of energy carriers for living systems?

Having formulated a general description of life, it is now necessary to try and fill out this description in terms of physical reality.

The first and most important question is what kind of elementary interactions or elementary forces can be responsible for the processes occuring in a living organism. We know that there are four elementary forces:

- strong (nuclear),
- electromagnetic,
- weak,
- gravitational.

The problem of which is best fitted to participating in living processes can be dealt with as follows. In our description of living systems the requirement for order has been emphasised. From this point of view, the four elementary forces differ very strongly from each other.

We recall that forces express themselves both by attraction and repulsion. Of the elementary forces, however, while all exhibit attraction, only one, the electromagnetic, appears as a repulsive force also.

Using forces of attraction, only one class of material order can be built, that is, a spherical body. The degree of order in a homogeneous sphere is very low: the change in position of the components within the sphere does not change the properties of the sphere. A really highly ordered system can only be built with the aid of the electromagnetic interaction in which the forces of attraction match those of repulsion. The answer, therefore, is surprisingly simple – the only elementary interaction which can be responsible for highly ordered systems is electromagnetism.

4.5.4 What kind of elementary particles can play the role of carriers of life?

The next question concerns the elementary particles.

There are two rather simple criteria for the selection of the elementary particles best fitted to the role:

- the particles must be relatively stable because the time span of the living organism is close to the cosmic scale of time (that is, 15×10^9 years or 10^{17} seconds),
- the elementary particles must be charged electrically and/or magnetically in order to be coupled with the driving force for life, the electromagnetic interaction.

The answer to the selection criteria is as follows. Only the nucleon (being a doublet of the proton and neutron, which belong to the class of baryons), the electron (which belongs to the leptons) and the photon, being the carrier of electromagnetic forces, can be considered as carriers of life. The electromagnetic interaction of nucleons, electrons, and photons occurs not only between the simple particles but also between the complex bodies built of these particles (molecules).

What kind of complex bodies can be built? The answer is:

- Nucleons (protons and neutrons) join together and form the atomic nuclei.
- Electrons and atomic nuclei join together and form atoms.
- Atoms join together and produce molecules.

All these cases have only one elementary force in common: the electromagnetic force.

The general answer then is that life, being a highly ordered and complex system, can be built only of atomic nuclei and electrons, which interact with the aid of photons.

4.6 The Chemical Elements, Particularly the Light Elements, Are the Carriers of Life

4.6.1 Why are the light elements best fitted for this role?

If our model of life is justified, then we can claim, as above, that the most suitable building blocks are the chemical elements, since chemical forces are no more and no less than a manifestation of the electromagnetic interaction. The logical consequence of this prompts the next question: Which chemical elements are the building blocks of life? The following points lead to the answer.

a) The cosmic abundance of the most suited elements must be very large. We know that the abundance of elements varies over 10–12 orders of magnitude (Figure 4.11). Only nine elements have cosmic abundances higher than 100 ppM (parts per million) atoms: hydrogen (H), helium (He), carbon (C), nitrogen (N), oxygen (O), neon (Ne), magnesium (Mg), silicon (Si), and iron (Fe). Only a few further elements show cosmic abundances greater than 2 ppM: sodium (Na), aluminium (Al), phosphorus (P), sulphur (S), chlorine (Cl), argon (Ar), calcium (Ca), and nickel (Ni). All the other elements (more than 70) have abundances below 2 ppM.

b) Some of the elements listed above – He, Ne, Ar – belong to the group of noble gases. They show practically negligible chemical activity and could therefore be excluded from the list of candidates for the chemical elements forming living systems.

c) Not all chemical elements form the same type of chemical bonds. The following is a simplified list of the types of chemical bonds and the elements with which they are associated:
  - covalent bonding – all chemical elements can use this simple bond;

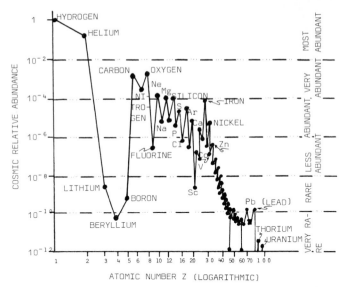

Figure 4.11. Only the light elements are abundant enough.

- covalent double bonding – only a small number of elements form bonds of this type, namely oxygen, carbon, nitrogen, and sulphur;
- ionic bonding – nearly all elements can bond in this way;
- hydrogen bonding – this is a rather special type in which not only the electrons but the nucleus of the hydrogen atom, the proton itself, participates. This type of bonding occurs with hydrogen and the elements oxygen, nitrogen, sulphur, carbon, fluorine, and to some extent chlorine;
- van der Waal's bonding, which results from the interaction of dipole moments and includes all chemical elements.

The consequence of such differentiated chemical properties of the elements can be formulated as follows. The living organism is probably of such complexity that only the application of all possible kinds of chemical forces would allow such a system to be constructed and to operate. Thus the most suitable elements are those which fulfill the "criteria" of high cosmic abundance and widely differentiated chemical properties. Only a few elements fulfil these criteria; in order of decreasing abundance these are:

hydrogen	H
oxygen	O
carbon	C
nitrogen	N
sulphur	S
and a few others	(P, Na, Cl, K, Fe, Mg, ...).

4.6.2 Why is hydrogen oxide – water – the unique medium for living organisms?

The two most abundant elements in the Universe, leaving out the chemically inactive helium, are the two elements at the top of the list of those elements most active in the make-up of living beings.

We have seen that the most abundant compound in the Universe is hydrogen oxide – water. All cold bodies, gas and dust clouds, comets, meteorites, and some planets contain water in larger or smaller amounts.

But water shows some very peculiar properties which make it a unique chemical compound in the Universe. Firstly, in the liquid state it retains its crystalline structure thanks to the numerous hydrogen bonds existing between the molecules. Secondly, liquid water has a higher density than solid ice. The reason for this is shown in Figure 4.12.

But why should these peculiarities have significance for our general discussion on life? Life is a highly ordered system. The best medium for the existence of such a system would be a crystalline structure. But a living organism needs an exchange with the environment and must transport both energy and matter internally. The speed of transport is connected to the physical state of the medium. In the gaseous state transport is quickest; in the liquid state it is slower by a factor of 10,000 or more and in the solid state by

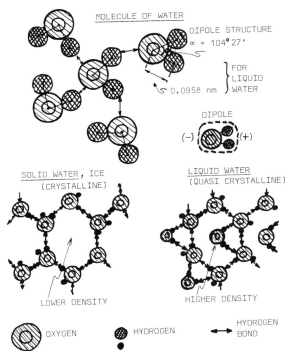

Figure 4.12. Water: hydrogen oxide. The peculiar, unique, and most abundant compound in the Universe.

even more, again by a further factor of 10,000–100,000. A living organism requires two things in this respect: a rather fast transport medium and an ordered crystalline structure of the medium. Both requirements are met by only one medium, water, at temperatures not far from its melting point.

The use by a living organism of water as its internal medium helps in several other important aspects:

- Water is an excellent solvent both for inorganic salts and for "organic" substances.
- Water remains a liquid over a relatively wide temperature range, about 100 K at a pressure of one bar and even greater temperatures at higher pressures. This guarantees the stability of the medium despite wide temperature fluctuations.
- Water dissociates into free hydrogen and oxygen under the action of an input of free energy (3 eV), such as ultraviolet light or 2 photons of visible light (each of ~ 1.5 eV).
- Free hydrogen and oxygen react spontaneously and intensively, resulting again in the synthesis of water. This occurs even at low temperatures and relatively low pressures.

We can see from these observations that there is no better medium for carrying life in the Universe than water.

4.6.3 The source of free energy for life: The stars of the Main Sequence

The internal and external medium for life is water. But life requires the continuous input of ordered energy or, as it is known in physics, free energy. Of course, the input flux of free energy presupposes the existence of a sink for the rejected energy, for the disordered energy, and for the bonded energy.

The characteristics required of a source of free energy are directly connected with the properties of the living system and its internal medium, water. These are some of the criteria:

a) The quality of energy must be high enough to be used to drive the chemical processes occurring in the living matter. The most vital chemical reactions are the transformations of free energy into energy stored in the chemical molecules. Since the most abundant molecule in the organism by far is water, we must evaluate the limitations of energy storage in the molecules of water. The breakup of the water molecule into hydrogen and oxygen needs an input of free energy of 237 kJ/mole, that is, 2.46 eV/molecule.

We know that a black body which emits at an energy of 2.5 eV must have a rather high temperature of some thousands kelvin. This energy corresponds to visible light, the red (wavelength ~ 800 nm) to photons with 1.55 eV, and the violet (wavelength ~ 400 nm) to photons with 3.1 eV. The free energy needed to decompose water (2.46 eV) corresponds to a photon

4.6 The Chemical Elements, Particularly the Light Elements, Are the Carriers of Life

with a wavelength of ~ 500 nm. The question remains, what energy source can be considered for this purpose – a blackbody with a temperature of some thousand kelvin. This could be a star.

b) The source of free energy must be long-lived, on the order of some billions of years ($>10^9$ years), because it is apparent that the chemical and biotic evolution of matter, including the evolution of a living system, needs such a period of time.

c) The flux of energy must be of the order of watts per square metre on the living organism which can make use of this free energy.

If we combine all these conditions for the existence of life with the properties of stars, then we obtain Figure 4.13.

It can be seen that the Main Sequence stars with masses more than 10 times smaller or 10 times greater than the mass of the Sun differ too much in temperature and therefore in the mean energy of the emitted photons. But the most dramatic difference occurs in the luminosity and lifetime. The stars with a mass 10 times greater than the Sun have (in relation to the Sun) the following properties: surface temperature 20,000 K (Sun, ~ 6000 K), photon mean energy 5 eV (Sun, 2 eV), luminosity a thousand times greater than that of the Sun and, consequently, a lifetime shorter than 100 million years. It seems that these stars cannot play the role of the free energy source for living matter, simply because their lifetimes are too short to give an opportunity for chemical and biotic evolution.

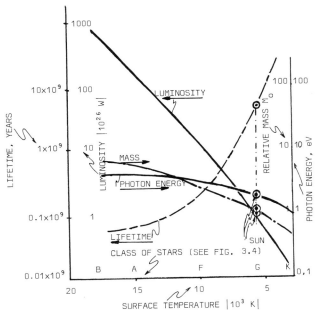

Figure 4.13. What type of star is most fitted to supply the free energy required by a living organism?

4.6.4 The chemical composition of the living organism is similar to the chemical composition of the Universe

There are a very large number of different types of living organisms on this planet. It is very difficult to give a definitive statement on the chemical composition of the biosphere, but the following gives a reasonable idea of the mean chemical composition of terrestrial life forms.

Atomic no.	Element	Percentage of atoms in nonliving biomass	Percentage of atoms in living biomass (66% H_2O)
1	Hydrogen	50.0	61.9
8	Oxygen	25.0	30.8
6	Carbon	24.0	6.9
7	Nitrogen	0.27	
15	Phosphorus	0.03	
16	Sulphur	0.017	0.4
—	All others	0.68	
	Total	100.00	100.00

Figure 4.14 also gives the relative chemical composition of a living terrestrial creature of the Earth's crust, of the oceans, and of the Universe (or rather the Solar System) (see Table 4.4 and Fig. 4.15).

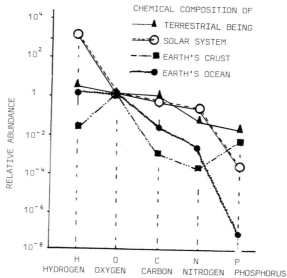

Figure 4.14. The chemical composition of the living being is similar to the chemical composition of the Universe.

4.6 The Chemical Elements, Particularly the Light Elements, Are the Carriers of Life

Table 4.4. General Characteristics of Life

Number	Characteristic	Elementary chemical carrier	Molecular form of carriers	Basis for selection
1	Spontaneous evolution on the cosmic scale	Hydrogen (H)	Hydrogen (H_2)	Hydrogen: – Is the most abundant cosmic element (93 atoms per 100); – Is chemically very active; – Exhibits the weak chemical bond (hydrogen bond); – Carries a maximum free nuclear energy.
2	Order (low entropy content)	Oxygen (O)	Water (H_2O)	Oxygen: – Is the second most abundant element.[a] Water: – Is the most abundant cosmic compound; – Is chemically active; – Carries an internal order (semicrystalline structure); – Is an excellent solvent; – Is stable over wide limits.
3	Requires steady flux of free energy (to combat otherwise continuous increase of disorder)	Carbon (C)	Carbohydrates and analogues ($H_{12}O_6C_6$)	Carbon: – Is the third most abundant element in the Universe; – Due to high valency and small radius, is a constituent of many molecules; – The carbon-containing molecules are also highly organised and form macromolecules; – The compounds are generally stable in aqueous media.
4	Control over the flux of free energy and matter between the organisation and the environment	Nitrogen (N)	Amino acids ($H_xO_yC_zN$)	Nitrogen: – Is the fourth most abundant element in the Universe; – Can be involved with H, O, and C in numerous compounds which are stable in aqueous media; – Forms a large number of complex macro-molecules.
5	Able to multiply (self-copying or reproduction)	Phosphorus (P)	Nucleotides ($H_xO_yC_zN_qP$)	Phosphorus is probably a component of those macro-molecules which are carriers of genetic information.

[a] The chemically nonactive helium is ignored here.

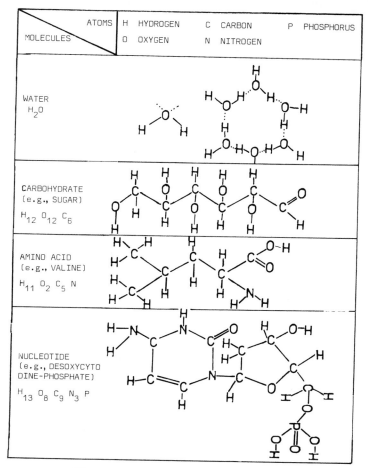

Figure 4.15. Chemical components of life.

4.6.5 Life is only possible in a Universe having the characteristics of our type of Universe

The existence of life in our Universe is at least one unarguable fact. We have discussed the problem of the origin of life based on the chemical interaction of the light elements in the neighbourhood of a middle-mass Main Sequence star which, over a period of some billions of years, evolves into a highly organised, thinking being.

Many other parameters were also touched on, such as the rate of evolution of stars, the rate of evolution of galaxies, and the evolution of matter from the primordial state in the Big Bang leading up to the synthesis of the heavy elements from a supernova explosion.

From these considerations we arrive at a principal question: How are all these subsequent evolutionary stages influenced by the evolution of the Universe itself? Putting it another way, how are we to understand the way in which the original starting properties of the Universe – the ground rules – have allowed evolution to pass through many complex and spontaneous stages to reach the most complex stage of all – the living being? Is life possible in another type of Universe with other starting properties, for example, a much higher mass density or another structural form? Of course, many ideas can be considered mere flights of fancy, since, by definition, the only Universe which exists is this Universe. In spite of this, philosophical discussions are useful and continue unabated.

One line of argument is that the presence of stars is possible only when galaxies exist, and that the formation and evolution of galaxies is possible only in a particular type of Universe.

Our Universe is isotropic and homogeneous, that is, it appears to have the same properties in all directions and in all parts, such as mean density of matter, rate of expansion, and mean temperature (the background radio waves associated with a blackbody temperature today of ~ 2.9 K).

These basic properties of the Universe are possible under the condition that the expansion velocity from the start was great enough to prevent an early gravitational collapse – that is, an "Anti-Big Bang" in the early Universe. On the other hand, the expansion velocity cannot be too large, otherwise the mixing of the components as they are formed to reach the neccessary isotropy and homogeneity is not possible. The theory states that the formation of galaxies and, therefore, stars is not possible in an anisotropic and heterogeneous Universe.

Only in a Universe in which the expansion is on the order of the speed of light is the isotropy and homogeneity and the formation of galaxies and stars possible, leading then to the formation of life and subsequent evolution to a thinking being.

Thus the existence of the higher forms of life is very closely coupled to the basic form and properties of the Universe, and that relationship may well be unique.

4.7 What Can We Hope to Know About the Spontaneous Formation of Terrestrial Life?

4.7.1 The problem: The uniqueness of life in our present state of knowledge

The greatest difficulty of any discussion about the origin of life arises from the fact that we know only one type of life, the terrestrial. We have not been able to observe any trace of extraterrestrial life. We are also unable to repeat the spontaneous formation of life from nonliving matter in our laboratories because of two constraints: firstly, the lack of a good theory and a proper experimental basis, and secondly, the very long time scales which would be needed for such an experiment. Failing these foundations we can only construct reasonably acceptable hypotheses which are, more often than not, guided by personal taste rather than pure scientific objectives.

4.7.2 The protobionts: The first living structures

Here we consider the Earth during the first step of evolution.

The so-called "primordial soup" contained a rather concentrated aqueous solution of ammonia, sulphur compounds, sugars, amino acids, nucleotides, phosphoric acid, and iron salts, and probably also some of the higher polymers, e.g., polypeptides and simple nucleic acids. These macro-molecules can play the role of space-organising membranes, and process rate controlling agents (catalysts and enzymes).

The first structures which began to transfer information about their own internal arrangement were formed spontaneously by a random process. The protobionts were able to take part in all processes necessary for life except one. Since the protobionts were separated from the "primordial soup" by a kind of permeable membrane, they exchanged matter with their environment mainly by consuming high-energy-content molecules and by emitting molecules with a low energy content. The rate of these processes was controlled by enzymes whose synthesis was rate-controlled in turn by the macro-molecules. However, the production of the energy-containing molecule adenosine-triphosphate (ATP) was not necessarily mastered by the protobionts, since this unit of energy currency was in any case present in the "primordial soup". With the help of this ATP, the living being can produce the hydrogen which is essential for almost all chemical processes of life. The oldest "fossil organisms" seem to be the "stromatolite carbonates", which are 3.5×10^9 years old.

4.7.3 The evolution of the living being occurred at the switch-over point from one energy source to the next

The unit of energy currency, ATP, which today plays the same role in the living organism, was formed in the primordial soup by spontaneous synthesis triggered by solar ultraviolet radiation. But in a short time the increasing population of protobionts, the primitive structures which are consumers of ATP, bring about a state where the rate of ATP production falls below the rate of ATP consumption. This, the first "starvation" in the history of living creatures, began to eliminate all living structures which consume ATP. However, some of the protobionts, thanks to the spontaneous changes in the

4.7 What Can We Hope to Know About the Spontaneous Formation of Terrestrial Life?

carriers of genetic information, possessed enzymes which were able to synthesise ATP from other components, that is, adenosine, ribose, and phosphoric acid.

This synthesis can only occur when free energy is added from outside. Such a source would be a chemical reaction in which small molecules containing oxygen, hydrogen, and carbon transfer some of the oxygen. This corresponds to an internal burning in the organisms known as fermentation. The scheme of fermentation is given as:

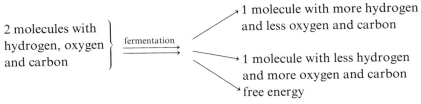

The hydrogen-rich molecule is the "fuel", the carrier of energy, and the oxygen-rich molecule is the "ash" or waste.

These protobionts, which have achieved by chance the advanced trick of synthesising ATP from other components by utilising the fermentation process, can survive and reach the next stage of evolution, the "eobionts".

The eobionts consume the carbon-containing molecules of the primordial soup. The concentration of these molecules, however, was small and decreased with time because the rate of production was smaller than the rate of consumption by the eobionts. Some of the eobionts overcame this by mastering the trick of consuming other eobionts, perhaps smaller ones or those with weaker outer membranes. These eobionts are the beginning of a line of typical heterotrophic organisms, that is, organisms which consume other, or at least some of the products of other, organisms.

The day comes, however, when the "primordial soup" is so depleted that no more energy-containing molecules are present. This is the second large starvation. To overcome this barrier life must have discovered and mastered a very sophisticated biological mechanism – the photosynthetic process.

The depleted primordial soup no longer plays the role of a source of free energy. The really giant source of energy is the flux of light coming from the central star, the Sun. The total energy flux in the solar radiation on the Earth correponds to 10^{17} W: that is, about 200–300 W/m² of the Earth's surface.

The simplest way to manage this is to use the solar light for the destruction of hydrogen-containing molecules, for example:

$$\text{hydrogen-containing molecule} \xrightarrow{+ \text{photon}} (\text{free hydrogen}) + \begin{cases} \text{waste} \\ \text{molecule} \end{cases}$$

The hydrogen-containing molecules that can be used are hydrogen sulphide (H_2S) and methane (CH_4). Unfortunately, the most abundant hydrogen-containing molecule, water (H_2O), is so strongly bonded that only a very strong ultraviolet photon is able to release hydrogen. Ultraviolet radiation, however, can release hydrogen equally well from the water molecules making up the essential life-carrying fluid of the living organisms themselves, thus killing them. The direct utilisation of ultraviolet radiation by living organisms must, therefore, not take place if the organism is to survive. Another way must be found.

At the stage when, in turn, the quantity of hydrogen-containing substances in the environment decreased significantly (the third starvation event) a further two-step mechanism became necessary: two-photon photolysis of water, occuring as:

$$\begin{pmatrix}\text{hydrogen oxide}\\ \text{water}\end{pmatrix} \xrightarrow[\sim 1.5\,\text{eV}]{+\text{first photon}} \begin{pmatrix}\text{molecule containing hydrogen}\\ \text{in excited state}\end{pmatrix} \rightarrow$$

$$\left.\begin{matrix}\text{molecule containing}\\ \text{hydrogen in excited state}\end{matrix}\right\} \xrightarrow[\sim 1.5\,\text{eV}]{+\text{second photon}} \begin{matrix}\text{free hydrogen}\\ \text{free oxygen}\end{matrix}$$

Both photons are from the region of visible light, that is, of energy less than 3 eV. Mastering this two-step photolysis of water allows the use of the radiation of the central stars of class G (see Figure 4.16).

This is the foundation for the existence and evolution of a biosphere over billions of years.

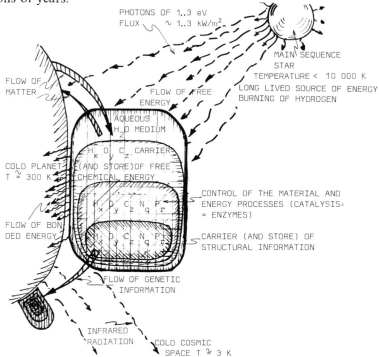

Figure 4.16. Model of the living being with the sun as a source of free energy (see also Table 4.4).

4.8 Evolution of Living Beings

4.8.1 Genetic evolution

To begin this section, we give one general remark: This book is devoted to the problems of the evolution of matter and energy, but we cannot ignore the fact that the description of the quantity and quality of all objects in Nature, of matter and energy, is given by the specialized language called "information".

The living being is not only the product of the intimate coupling of matter and the flow of free energy, but is also a product of the ancient flow and evolution of the information about this coupling process, the genetic information.

We begin with some rudimentary information about the information itself.

In a very simplified calculation of the information content of primordial spontaneously produced peptides in the primordial soup, let us arbitrarily postulate the following numerical values:

$$(2 \times 10^{18} \text{ kg}) \times (5 \times 10^{25} \text{ atoms/kg}) = 10^{44} \text{ atoms}.$$

The number of amino acid molecules which could be made from these atoms of carbon: (only $1:100$) = 10^{42} molecules.

The number of primitive peptides, being polymers with only 10 amino acid molecules, equals $\sim 10^{41}$ macro-molecules. The mean "life" of peptides, up to the moment of spontaneous decay, equals approx. 10^{-2} second.

Duration of the "primordial soup" era equals:

$$500 \times 10^6 \text{ years} = 1.6 \times 10^{16} \text{ s}.$$

Total number of possible molecular combinations of peptides:

$$N = 10^{41} \times (1.6 \times 10^{16})/10^{-2} = 1.6 \times 10^{59}.$$

The quantity of "information" (I) in the primordial soup equals:

$$I_{\text{Earth}} = \log_2 N = \log_2 (1.6 \times 10^{59}) = 197 \text{ bits}.$$

Conclusion: the total information content of primordial soup during half a billion years has been calculated as equal to only 197 bits.

We can now calculate the quantity of information (I) of all primordial soups in the whole Galaxy with an arbitrarily taken number of "habitable" planets with an equal potential for evolution of life:

Number of "habitable" planets: $\sim 10^9$ (see Chapter 9.7)

$$I_{\text{Galaxies}} = \log_2 (1.6 \times 10^{59} \times 10^9) = 227 \text{ bits}.$$

Even this value of information is rather low, but the number of molecular combinations is astonishingly large.

Below we give a small table with the information contents of various living beings without (at present) further explanation.

Living being	Number of possible cases of different states N	Amount of information I (bit)	Number of genes (carrier of genetic information)	Number of "letters" in the genetic code
Virus	$10^{36,000}$	120,000	50	4×10^4
Escherichia Coli (Bacterium)	$10^{1,300,000}$	6,000,000	2500	4×10^6
Man	$10^{72,000,000}$	240,000,000	100,000	3×10^9

Compare the number of possible different states of a human body with the number of atoms in the Universe. Do not forget that the total number of atoms in the Universe equals only:

$$N_{\text{Universe}} = 10^{80} \text{ atoms}.$$

4.8.2 The evolution of Man

The highest and most ordered structure in the Universe, at least according to our present state of knowledge, is Man. The evolution and the origin of Man is, at the present time, more or less clear (Figure 4.17).

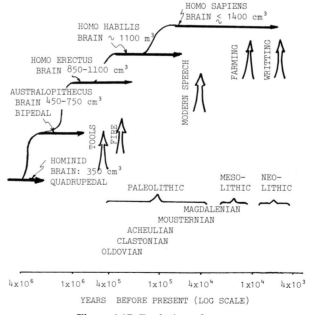

Figure 4.17. Evolution of man.

4.8 Evolution of Living Beings

A particularly interesting problem is the evolution of the size of the brain (Figure 4.18). But this does not explain the uniquenes of Man and the surprising possibilities of his intellect.

4.8.3 The evolution of the brain

From the biological point of view, the most specific characteristic of Man is the extremely high relationship between the mass of his brain and the mass of his body, or more exactly, the so-called "encephalisation quotient" EQ, which equals (in kg):

$EQ = 1.2 \times P^{(2/3)}/B$;
P = mass of body (kg);
B = mass of brain (kg)   (Man's brain: average 1.3 kg);
$EQ_{Man} = 1.2 \times 60^{(2/3)}/(1.3) \cong 12$

(see Figure 4.18).

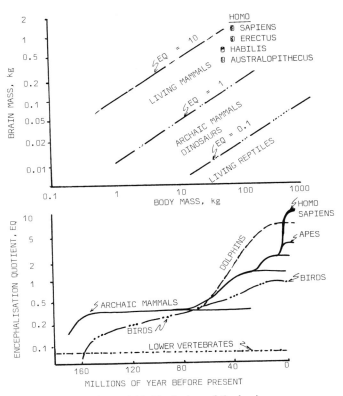

Figure 4.18. Evolution of the brain.

In the evolution of animals, Man, being the youngest branch, achieves the highest encephalisation quotient. The question is, will this process continue or is further encephalisation not possible? This problem is not a purely academic question, and will be discussed in Chapter 9.

4.9 What Are the Conclusions for the Future of Mankind?

Based on the points raised in this chapter:

- the chemical processes in cosmic space, the dust-gas clouds, comets, meteorites, etc;
- the model of the origin and evolution of the planets;
- the chemical evolution of the Earth's primitive atmosphere;
- the general definition of a living being;
- the models of the spontaneous origin of the terrestrial biosphere;
- the connection between living beings and the sources of free energy and the evolution of information content,

the following conclusions can be drawn:

- Planetary chemical evolution and the spontaneous creation of life has a low probability and may be quite rare in the Universe.
- The probability of finding extraterrestrial life forms is small but certainly not zero.
- The evolution of the biosphere is heavily dependent on the sources of free energy. This gives a basis for prediction of the subsequent evolution of life over the next billion years.

The decisive choices still lie with Man himself.

CHAPTER 5

The Eternal Cycle of Matter on the Earth

> Two conflicting views dominate our perception of man's long-term future. The "catastrophists" believe that the Earth's resources will soon be exhausted ... the "cornucopians" argue that most of the essential raw materials are in infinite supply.
>
> A.M. Weinberg
> (1915-   )

> All things from eternity are of a like form and come round in a circle.
>
> Marcus Aurelius
> ("Meditations")
> (121–180)

> Don't drink the water and don't breathe the air.
>
> T. Lehrer
> ("Pollution", 1965, from P. Ehrlich "Ecoscience")

5.1 Matter on This Planet Is Almost Indestructible

5.1.1 How stable is terrestrial matter?

The first questions to ask are, what kind of matter are we considering and what do we mean by stability?

It must be noted right at the start that, for at least four billion years, terrestrial matter has existed under rather soft or easy conditions, far from extreme values. For instance, the terrestrial gravitational field is much weaker than that of any star, especially the very dense ones such as neutron stars or black holes. However, the field is not as weak on Earth as that prevailing in dust-gas clouds; similarly for temperature, not as high as in the stars and not as low as in the interstellar clouds.

Again the age of the planet is not as great as the age of the galaxies but is long in relation to, for example, the half-life of many radionuclides.

We know that under these "soft" conditions only a small number of massive elementary particles are stable. These are:

- the hadron class – subclass of the baryons: the proton and the bonded neutron;
- the lepton class: the electron.

These stable baryons, of course, form the atomic nuclei and then, together with the stable leptons, the electrons, form the atoms.

The stability of these elementary particles is also covered by the principle of the conservation of baryon number and lepton number. It is "forbidden" to alter the sum of all (koino- and anti-)baryons and, similarly, the sum of (koino- and anti-)leptons.

5.1.2 Terrestrial matter is isolated by the gravitational field; the amount of matter is constant

The amount of terrestrial matter is held constant by the gravitational field of the Earth.

The total amount of material is some 6×10^{24} kg, mostly in solid form with a mean density of 5500 kg/m^3, which is large enough to form a gravitational field strong enough to prevent all but a few particles, including single atoms, from escaping from the Earth. To do so they must achieve a velocity of ~ 11 km/s, and under normal conditions only the light particles can achieve this. For example, much hydrogen and helium in the upper layers of the atmosphere, where the pressure and the chance of interatomic collisions are low, achieve the escape velocity and leave the Earth forever.

Other than these chance escapes, only artificial objects – space rockets – can be sent by man far from the gravitational field of the Earth (Figure 5.1). The high price which must be paid in energy terms to separate these rockets from the Earth is well-known, and it seems quite certain that in the future the transport of some materials from other planets to Earth will be similarly prohibitive. This is especially true when it is remembered that all the stable and quasi-stable elements already exist on this planet in rather large amounts.

5.1.3 Division of the Earth into five "spheres"

The first differentiation in the original homogenous mixture of elements held in dust particles and gas occurred in the very early phase of the protoplanet's formation.

Today it is possible to define the following divisions:

5.1 Matter on this Planet Is Almost Indestructible

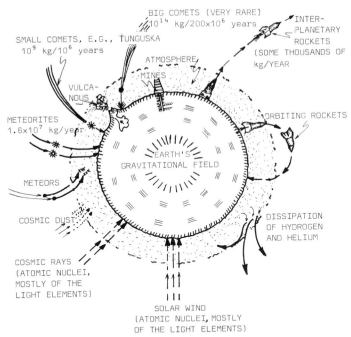

Figure 5.1. Most terrestrial matter is held "forever" by the gravitational field of the Earth.

a) lithosphere: the solid sphere which contains 99.977 weight-percent of the terrestrial material;

b) hydrosphere: the sphere where matter is in liquid form or dissolved, containing 0.023 weight-percent, or 234 ppM, of the total terrestrial matter;

c) atmosphere: the sphere of gaseous material consisting of 0.85 ppM of terrestrial matter;

d) biosphere: the sphere of highly organised matter, mostly in a macro-molecular polymerised state, made up of 0.83 parts per 10^9 (ppG) of terrestrial matter;

e) technosphere: the sphere of man-made objects, mostly in solid form, making up an estimated 1 part per 10^{12} (1 ppT) of all matter on the Earth (see Figure 5.2).

The chemical composition of each of these spheres is different from the others, but at least one element is common to all, namely oxygen. Oxygen provides half the mass of the lithosphere, the bulk of the hydrosphere, a fifth of the atmosphere, a quarter of the biosphere, and probably a quarter of the technosphere. But for us in this chapter, not just the total amount of terrestrial matter is of interest. Much more significant is the rate at which the material is

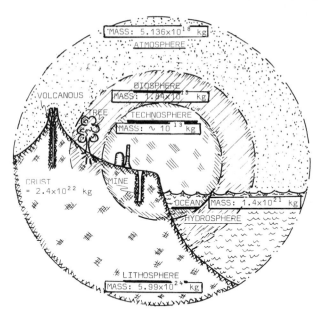

Figure 5.2. The five terrestrial spheres – lithosphere, hydrosphere, atmosphere, biosphere, and technosphere – are interconnected.

cycled. This data is shown in rather rough form in Figure 5.3 together with the masses (absolute and relative) of each of the five spheres.

It is impressive to see that at present the material cycle of our civilisation, including water, foods, fuels, etc., has reached a level which is of global significance.

Now we turn to a more systematic review of the material cycle in each of the five spheres.

5.2. The Gaseous Sphere Acts in the Exchange Between the Other Spheres

5.2.1 The main components of the atmosphere

The reasons why we begin reviewing the properties of the terrestrial spheres with the atmosphere are:

a) The gaseous state permits the fastest transfer of matter due to high diffusion rates and atmospheric mass transfer such as turbulent air streams and winds.
b) One of the main components, oxygen, is one of the most active chemical elements, particularly in the biosphere.

5.2 The Gaseous Sphere Acts in the Exchange Between the Other Spheres

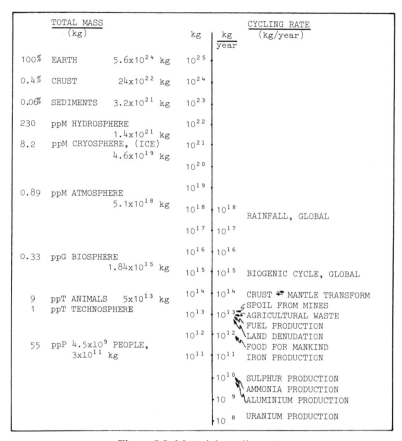

Figure 5.3. Materials cycling rate.

c) Water in the given temperature range is rather volatile and can be transported over global distances.

d) The atmosphere absorbs some of the solar energy shielding the Earth's biological organisms from harmful solar radiation.

e) The atmosphere is the most important carrier of heat energy in two directions – horizontally across the continents and oceans and vertically from the surface to cosmic space.

The chemical composition of the atmosphere is rather simple, especially in relation to the other spheres. One of the components, water vapour, is present in varying amounts (Figure 5.4).

The atmosphere contains the elements most important to the role of carriers of life – hydrogen (in the form of water), oxygen (in water and as free molecules), carbon (in carbon dioxide), and nitrogen (in a free molecular form).

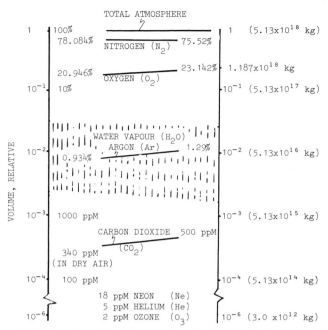

Figure 5.4. The atmosphere has a rather simple chemical composition.

5.2.2 The most active component, oxygen, a product of the biosphere

Free oxygen is an extremely active element and reacts with almost all other elements, forming many chemical compounds. How can such an active element be so prominent in the atmosphere? Why is it not all removed by chemical reactions? Before coming to the answer, we present some details of the oxygen levels in the atmosphere.

The total amount of free oxygen at present is 1.18×10^{18} kg, which is equivalent to 20.96 mol percent of the total. The relative error does not exceed 0.0017 mol percent, that is, approx. one in ten thousand. This estimated oxygen content has remained stable for measurements made over the last 70 years in spite of the increase in the amounts of fuel – oil and coal – burned in the technosphere! In addition to this technosphere activity, other chemical processes are occurring in the lithosphere and hydrosphere, for example (Figure 5.5):

a) oxidation of iron oxide in ores from ferrous (Fe^{+2}) to ferric (Fe^{+3}) oxide:
    $FeO + \frac{1}{2}O_2 \rightarrow FeO_{3/2}$;
b) oxidation of elemental sulphur (S^0) to the sulphate ion (SO_4^{-2}):
    $S^0 + 2O_2 + 2e^- \rightarrow SO_4^{-2}$);
c) burning of hydrogen expelled by volcanoes: $H_2 + \frac{1}{2}O_2 \rightarrow H_2O$.

These processes are oxygen-consuming. But there are also oxygen-producing processes, such as those occurring in the upper layers of the atmosphere:

5.2 The Gaseous Sphere Acts in the Exchange Between the Other Spheres   111

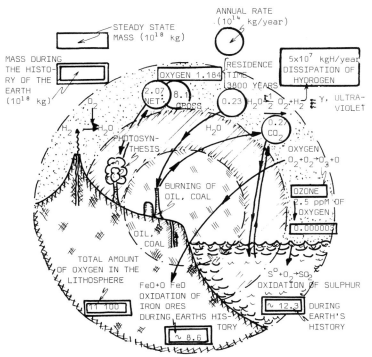

Figure 5.5. The oxygen cycle in simplified form.

d) photolysis of water molecules by solar ultraviolet radiation connected with the dissipation of free hydrogen; oxygen remains in the atmosphere:

$$\text{water} \xrightarrow{\text{solar ultraviolet}} \text{hydrogen} + \text{oxygen}$$

                                             dissipated   remains in
                                             into space   atmosphere

Without doubt the most significant process in the oxygen cycle is photosynthesis, which can be thought of as photolysis in which the free hydrogen is bonded to carbon dioxide, resulting in the formation of formaldehyde:

$$\text{water} \xrightarrow{\text{solar light}} \text{hydrogen} + \text{oxygen}$$

                                             bonded by carbon   remains in
                                           dioxide to form   atmosphere
                                           formaldehyde
                                           $(2H_2 + CO_2 \rightarrow H_2CO + H_2O)$

The amount of oxygen produced by photosynthesis equals $\sim 3.1 \times 10^{14}$ kg/year gross and $\sim 2.07 \times 10^{14}$ kg/year net. This corresponds to an annual net production of 1.725×10^{14} kg of dry organic matter. The flux of net absorbed solar energy for photosynthesis is 92 TW or 2.9×10^{21} J/year: gross energy equals 136 TW or 4.3×10^{21} J/year.

This annual production rate of oxygen is equivalent to $\sim 2.6 \times 10^{-4}$ parts per year of the total atmospheric oxygen, or a full recycling time of $\sim 3,800$ years.

The technological processes of mankind require at present approx. 10 TW, corresponding to an amount of oil and coal equivalent to 7.5×10^{12} kg annually. The amount of oxygen required is approx. 3.1 times the amount of carbon and hydrogen used in the fuels, which is equivalent to 2.3×10^{13} kg oxygen/year.

We see, then, that the amount of oxygen used in the technosphere in relation to the net production of oxygen in the biosphere is about 11%. Can this influence the global balance of oxygen? What is the global balance of carbon dioxide?

It is important to realise that compared to the total amount of oxygen in the atmosphere, some 1.18×10^{18} kg, the annual rate of consumption of oxygen in the technosphere is only 19 ppM. At the present rate, only after 50,000 years will all the atmospheric oxygen have been "burned" by technological processes.

Our measurement of the concentration of oxygen in the atmosphere goes back 70 years. In this time, the total calculated amount of oxygen "used" is approx. 300 ppM of the total. The accuracy of the present measurement is about 80 ppM, but no changes in the oxygen concentration in the atmosphere have been observed. Probably after a further decade we will be able to answer the question of how much the technosphere has altered the balance of free oxygen due to the burning of fossil fuel.

5.2.3 Ozone: Modified oxygen which acts as a shield for the biosphere

Free oxygen is one of the substances vital for the biosphere, at least in its present state. It is the carrier of free energy which supports high-energy-demanding metabolic processes in living beings. The chemical form is the diatomic molecule O_2.

However, the triatomic molecule of oxygen, O_3, called ozone, plays a surprisingly important role in sustaining life on the land masses and in shallow water, where 99% of living creatures exist. Ozone is produced by the action of ultraviolet solar radiation with a wavelength less than 242 nm, that is, with an energy higher then 5.3 eV/photon:

molecular oxygen + molecular oxygen $\xrightarrow{\text{ultraviolet radiation}}$ ozone + atomic oxygen.

The amount of ozone in the atmosphere is more or less constant, because in parallel to this synthesis of ozone there exist also numerous reactions which destroy the ozone, once formed:

Ozone O_3
- → molecular oxygen (18%) → $O_2 + O_2$
- → hydroxyl radical (11%) → $O_2 + H$
- → nitrogen monoxide (50–70%) → $O_2 + NO_2$
- → unknown compounds (0–20%) → $O_2 + O_2$

The percentages represent the proportion of each reaction to the total of the ozone destroying processes.

It appears that for hundreds of millions of years both types of processes, the formation of ozone by ultraviolet radiation and the destruction of ozone, have been in equilibrium. The steady-state concentration of ozone in the atmosphere is 0.6 ppM, that is, ~2.5 ppM of atmospheric oxygen. If all this were compressed into a gaseous layer with a pressure of 1 bar (the mean atmosphere pressure at sea level), it would form a layer only 3 mm thick. The total mass of ozone equals approx. 3×10^{12} kg.

The importance of such a small amount of ozone lies in its ability to absorb the ultraviolet part of solar radiation in the higher regions of the atmosphere. This ultraviolet radiation, with 5 eV energy per photon, is able to destroy water molecules, that is, to cause, in a single step, a very effective photolysis of water. In a living organism, which contains from 60–95% water, the direct action of ultraviolet radiation results in very rapid death. In smaller doses it can also result in the initiation of skin cancer in higher organisms. Some calculations show that a decrease in the ozone layer of ~5% could increase the dose of ultraviolet radiation on the land by about 10%. This can cause about 1–3 additional cancers per thousand of the human population, particularly in the light-skinned races. (Fortunately, these are mainly skin cancers which appear to be treatable.) The influence of an increase in ultraviolet radiation on green plants is not clear. Similarly, the climatic repercussions are not clear but may be significant.

All such effects are only of significance over a long time scale. Over a period of 24 hours the level of ozone may vary significantly. It seems that the concentration of ozone and the temperature of the ozone layer vary in line with fluctuations in solar activity. The influence of the products of the technosphere will be discussed later.

5.2.4 The carbon cycle, a chain directly related to the flow of energy in the biosphere and technosphere

Going on from the discussion of the part played by oxygen in the atmosphere, it is only a short step to the role of the carbon cycle, which is directly coupled to the behaviour of oxygen.

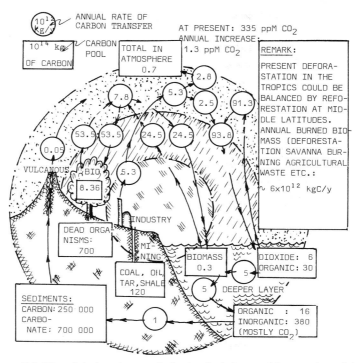

Figure 5.6. The global carbon cycle is clearly influenced by man's activity.

At the present time the atmosphere contains about 327 ppM of carbon dioxide by volume, or 500 ppM by weight, which corresponds to a total of about 7×10^{14} kg of carbon (Figure 5.6). The variation in carbon concentration from the mean value of 335 ppM ranges from 400 ppM near the Earth's surface at night to approx. 300 ppM at midday in thick forests due to photosynthesis.

The carbon concentration is the result of the photosynthesis which has been occurring over the last billion years. However, during the last century, man has made a large impact, with his destruction of forests and burning of coal and oil, the fossil reserves of carbon and hydrogen, which appears to have upset the "eternal" steady-state cycling of carbon on this planet. Some calculations show that in the hundred years from 1850 to 1950 a total of 120×10^{12} kg of carbon was released into the atmosphere from deforestation and burning dead organic material, while over the same period the increase from burning fossil fuels was half this value, that is, approx. 60×10^{12} kg of carbon.

At present man burns about 5.3 billion tons of carbon-containing fuels annually (that is, approx. 5.3×10^{12} kg/year) and causes the destruction of wood and humus (deforestation) amounting to a further 2×10^{12} kg/year. The total carbon release to the atmosphere equals about 7.5×10^{12} kg/year

5.2 The Gaseous Sphere Acts in the Exchange Between the Other Spheres

(Figure 5.6). This amount, compared to the available carbon in the atmosphere, is:

$$\frac{7 \times 10^{14} \text{ kg C in atmosphere}}{7.5 \times 10^{12} \text{ kg C per year}} = \text{equivalent of } 100 \text{ years}.$$

This means that in a little over 100 years of fossil fuel burning the amount of carbon in the atmosphere will have doubled, increasing the concentration to 640 ppM. In reality, the rate of increase of atmospheric carbon is ~ 5 times smaller, due to absorption in the oceans and the photosynthesis (by algae) in near-shore shallow water due to eutrophication and other as yet unknown processes. It is clear that such a doubling of carbon dioxide in the atmosphere, being a dramatic change in one of the most important features of the terrestrial atmosphere, even when moderated by the oceans, must give rise to some anxiety.

One of the most significant results of the increase in carbon dioxide levels is the reduction in the transparency of the atmosphere to infrared radiation. This radiation, emitted by the Earth's surface after heating by solar energy, is a critical feature of the heat balance of this globe (see Section 6.2.2, Figure 6.4). Some experts consider that the decrease in transparency to infrared radiation will become so severe that the mean global temperature could rise by two or three degrees, enough to cause melting of the polar ice and an increase of ocean levels. Others counter this argument with the view that the increase in mean temperature will increase the cloud cover and decrease the solar energy input, thus restoring the thermal balance. According to recent considerations, because of carbon dioxide's "greenhouse effect", the climate will become hotter, wetter, and cloudier. Doubling atmospheric carbon dioxide raises the global temperature by about $2°$ C, with the effect concentrated at the poles. But it must be emphasised that the cycling of carbon dioxide is enormously complicated and includes many important areas of ignorance. The capability of the oceans to absorb carbon dioxide shows signs of saturation, but this remains uncertain. Moreover the world's forests, which for so long have been considered a second "sink" for carbon dioxide, are beginning to shrink (see Chapter 7.3). Instead of being a sink, the forests have become a source of carbon dioxide as they are burned.

This situation seems to be the most important sign of Man's impact on the global scale. The solution needs important international decisions within the next decade. If the decisions are postponed, then for all practical purposes the die will already have been cast.

One thing is sure, namely Man's activity in this case has already reached such a level that the consequences could be felt on a global scale and be of dramatic significance. Before continuing, it is perhaps time to take stock and make a careful evaluation of the impact of all this activity on steady-state global processes.

The other important form of carbon in the atmosphere is carbon monoxide, CO. The burning of fossil fuels, especially in today's cars, is a source of about 3.5×10^{11} kg of carbon per year. How dangerous is carbon monoxide for the biosphere, for man? There is no argument that carbon monoxide is a lethal poison in relatively small concentrations. Carbon monoxide does not, however, accumulate in the terrestrial atmosphere.

5.2.5 The "inert" nitrogen cycle, which controls the activity of the biosphere

Although we will begin this discussion on the components of the atmosphere with oxygen and carbon dioxide, the biggest component of the atmosphere is, in fact, nitrogen. Approximately 79 out of every 100 molecules in the atmosphere – that is 75.5% by weight – are nitrogen.

The nitrogen exists in molecular form (N_2) and is a rather inert gas. It would appear to be merely a dilutant for the active agents of the atmosphere, oxygen, carbon dioxide, water vapour, etc., but this is not so.

Each living organism, at least on this planet, is made up of proteins, which play an essential role in the processes of life. The proteins are polymers, built from 20 different monomers, the amino acids (Figure 4.6). As the name suggests, each amino acid consists of at least one amino group, that is, a molecular group of one nitrogen atom bound with two hydrogen atoms – NH_2. There is no "life" without nitrogen.

The transmutation of the inert molecular nitrogen to the life-carrying amino acid groups is one of the greatest wonders of the biosphere. Each year approximately 70 billion kilograms of nitrogen are involved in biogenic processes. A further 8 billion kilograms are fixed by electrical discharges in the atmosphere.

Whereas in the case of the carbon cycle man's activity is influential in one direction only, that is, in the transfer of carbon from the Earth's crust to the atmosphere due to combustion of fossil fuels, in the case of the nitrogen cycle man's impact is in the other direction. The industrial fixation of nitrogen from the atmosphere by means of the synthesis of ammonia from atmospheric molecular nitrogen is presently running at levels of some 60 billion kilograms annually (Figure 5.7). This can be compared with the approx. 54 billion kilograms of nitrogen fixed by biological activity and a further 8 billion kilograms by electrical discharge. This technological contribution is unsurpassed in the other material cycles. Here man's activity is equal to that of the biosphere!

Of course, the positive side of this activity, the industrial fixation of nitrogen, is overwhelming, but there is probably an additional significant negative effect. In Chapter 7, both aspects are discussed: the benefits of the enhanced food production arising from the use of nitrogen fertilizers and the

5.2 The Gaseous Sphere Acts in the Exchange Between the Other Spheres

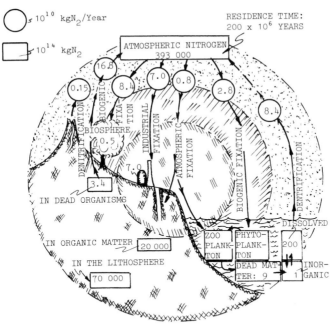

Figure 5.7. "Inert" nitrogen has a rather active cycle.

problem of the damage to the flora and fauna of lakes, rivers, and estuaries caused by the influx of these same fertilisers because of eutrophication. At present the balance is positive. How long will it remain so?

5.2.6 The micro-components of the atmosphere, the troublesome "details"

It is a familiar situation that it is not the principal elements or main components of a system which cause difficulties or influence its functioning, but the micro-elements, the details, which often give rise to concern, particularly where Man and the biosphere are concerned (Figure 5.8). The micro-components range from ozone (Chapter 5.2.3), consisting of ~ 1 ppM by weight, and sulphur dioxides with 1 ppG, to methyl iodine with 1 ppT.

Not only are the mean values of concentration significant, but also the annual production rates, whether from natural of artificial (man-made) processes and the corresponding depletion rates, all of which influence the role each component plays in the atmosphere. A parameter which links these factors together is the time of residence (t_{res}):

$$t_{res} \text{ (years)} = \frac{\text{amount in the atmosphere (kg)}}{\text{rate of production (kg/year)}};$$

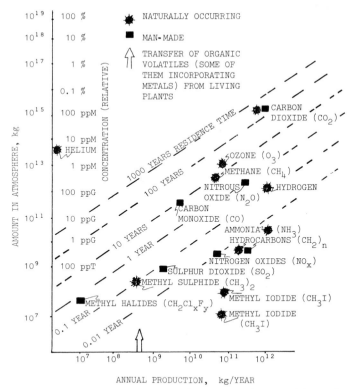

Figure 5.8. The micro-components of the atmosphere: residence time.

for example, sulphur dioxide:

$$t_{res} = \frac{10^9 \text{ kg}}{10^{11} \text{ kg/year}} = 0.01 \text{ year}.$$

We can see that the residence time fluctuates from 10,000 years for the main components to much less than 0.1 year for the sub-micro-components.

These trace compounds with a mean concentration of 1 ppM or less and with residence times of less than one year are unable to reach a homogeneous distribution in the atmosphere. Instead, their concentration may vary sharply – for example, in the vertical direction, as we have seen for ozone – but this is only one of many such cases.

As one of the better-known examples, consider the fluorocarbons. These micro-components are used in aerosol spray cannisters as a propellant and as a coolant in refrigerators. The very properties which make the fluorocarbons valuable as propellants, their inertness and insolubility in water, prevent them from being washed out of the lower atmosphere by rain. They gradually drift up to the stratosphere, where they are broken down by ultraviolet radiation, releasing free chlorine atoms. These atoms in turn react with and destroy ozone molecules, causing the Earth to lose its protective ozone envelope, the only protection we have from increased ultraviolet radiation – the ultimate consequence of which would be an increase of skin cancers in humans and additional dangers to other animals and to plants.

This hypothesis can be investigated by direct measurement of the chain of these reactions in the laboratory, in extremely low-pressure conditions, taking into account the influence of the walls of the reaction vessels. Some results show that the propellants are less harmful than was previously feared.

Some scientists believe that the fluorocarbons already released may reduce the ozone concentration rather significantly. Their calculations predict that the ozone layer will eventually be depleted by about 10% if fluorocarbon usage continues at its present level. The concentration of ozone at 40 km might decrease by $\sim 20\%$, when the concentration at 10 km could increase by 20%. Increasing concentrations of ozone in the lower atmosphere pose potential risk to air quality over the surface of the globe.

Here we have an example of the potential impact of an exclusively man-made volatile substance even though it makes up only a minute fraction of the atmosphere.

5.2.7 Dust particles, a troublesome constituent of the atmosphere

The atmosphere is gaseous. The main sub-components are gaseous, although some transform into liquid, for example, drops of water. Here, however, we deal with the solid matter in the atmosphere, the dust particles. These owe their long residence time in the atmosphere to their small size.

The concentration of dust particles in the atmosphere varies strongly with altitude, from 10 ppG in the lower levels to 1 ppT at altitudes of 3–15 km and increasing to 1 ppG in the stratosphere. Assuming that the mean concentration is 100 ppG, this means a total dust content of 5×10^{11} kg. In the stratosphere alone the amount is about 10^{10} kg. The amount in the lower atmosphere – made up of dust from deserts, forest fires, salt particles from oceans, industrial dust, and volcanic activity – varies over a wide range.

In the upper regions of the atmosphere the amount of dust is more constant, but may be strongly influenced by sudden volcanic activity. In the long term the influence of man's activity on dust production is also important, even in the stratosphere (Figure 5.9).

A dramatic impact would be made by a global nuclear war with approx. 10,000 nuclear explosions with a total yield of 5000 megatonnes. It is supposed that 10^{12} kg of dust would be created and 80% of it would find its way to the stratosphere, blocking sunlight. The consequence would be a "nuclear winter", which would severely inhibit photosynthesis.

The influence of dust particles on the Earth's heat balance by changing the albedo is discussed in Chapter 6.

Cosmic sources also account for very long-term changes in the total dust content of the atmosphere.

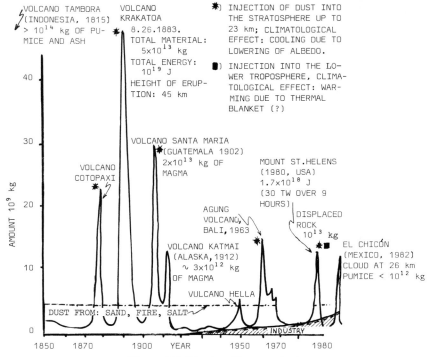

Figure 5.9. The troublesome dust.

5.3 The Hydrosphere – A Crucial Factor in the Existence of the Biosphere

5.3.1 The cycling of water, the largest terrestrial material cycle

Before going into detail on the make-up of the hydrosphere it may be claimed that, firstly, it is unique in the Solar System. Only the Earth has a hydrosphere, a liquid region formed from water. Of course other planets, particularly Jupiter and Saturn, probably have liquid zones but these are probably spheres of liquid ammonia and liquid hydrocarbons.

Secondly, water is also unique in being the most abundant chemical compound in the Universe; the terrestrial hydrosphere is a good representative of the Universe's average chemical compound.

Thirdly, a molecule of water has very specific and unique chemical and physical properties, arising out of the maximum number of chemical bonds per atom which can be formed (see Figure 4.12 and Section 4.6.2).

Fourthly, each living organism on the Earth is composed mainly of water – not less than 60% and in some cases 95% of the mass of an organism.

5.3 The Hydrosphere – A Crucial Factor in the Existence of the Biosphere

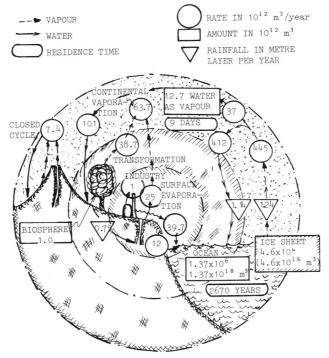

Figure 5.10. The giant water cycle.

Now the hydrosphere itself. The world's oceans cover about 71% of the Earth's surface, that is, 361.4 million square kilometres, or $3.61 \times 10^{14}\,\text{m}^2$. The mean depth of the oceans is 3790 metres. The volume of the oceans therefore equals 1.37×10^{18} cubic metres, with a mass of $1.37 \times 10^{21}\,\text{kg}$. The total amount of water on the Earth is, of course, greater, $1.46 \times 10^{21}\,\text{kg}$. A good part of this hydrosphere, in spite of the name, exists in solid form as ice (Figure 5.10). The ice sheet (cryosphere) amounts to some $4.5 \times 10^{19}\,\text{kg}$, that is, $\sim 4.5 \times 10^{16}\,\text{m}^3$, equivalent to a coating of ice 90 metres thick over the entire globe!

Not only liquid water, but also water vapour is important to life. Each year some 5.13×10^{14} cubic metres of water are vaporised by evaporation of liquid water. This is equivalent to a 1-metre layer of water over the entire globe. From the oceans alone about $4.45 \times 10^{14}\,\text{m}^3$ of water is transformed into vapour, a layer in the oceans of $1.14\,\text{m/year}$ supplying the continents $0.75 \times 10^{14}\,\text{m}^3$.

The total atmospheric water content is about $0.127 \times 10^{14}\,\text{m}^3$. The residence time ($t_{\text{res}}$) can be calculated as:

$$t_{\text{res}\,\text{atmosphere}} = \frac{0.127 \times 10^{14}\,\text{m}^3}{5.13 \times 10^{14}\,\text{m}^3/\text{year}} = 0.025 \text{ year or } \sim 9 \text{ days.}$$

The residence time of the water in the oceans is:

$$t_{res\,ocean} = \frac{1.37 \times 10^{18}\,m^3 \text{ in the ocean}}{5.13 \times 10^{14}\,m^3/\text{year}} = 2670 \text{ years}.$$

How much energy is needed to transfer this tremendous quantity of material by vaporisation? Roughly calculated the amount of heat required is given by (heat of vaporisation 2.5×10^9 J/m³)

$$H = (5.13 \times 10^{14}\,m^3/\text{year}) \times (2.5 \times 10^9\,J/m^3) = 1.28 \times 10^{24}\,J/\text{year}.$$

The energy flux is 1.28×10^{24} J/year, or 40.8×10^{15} watts, or 40,800 TW (Terawatt). When compared to the amount of solar energy absorbed by the Earth this is found to correspond to one-third.

From the point of view of the biosphere and of mankind, the main problem is the unequal distribution of rainfall on the Earth. It is known that large areas of the globe have a negative water balance. A major problem of mankind is that a significant part of the population lives in these areas. Specifically the need for water for agriculture seems to be particularly crucial (Chapter 7).

5.3.2 Quality of water, quality of life

Water is present in many areas apart from the hydrosphere proper; it is a very good solvent for many of the chemical compounds that make up the Earth's crust and atmosphere; it also makes up a large part of the biosphere.

The ocean is far from being a source of pure water and its chemical composition is very complex (Figure 5.11). The most abundant elements are the lighter ones, the heavier elements making up less than 1 ppM. One surprising feature is that carbon is present only at levels of 30 ppM.

Even more surprising is the fact that the total marine biosphere, with a dry mass of $\sim 4 \times 10^{12}$ kg (Chapter 7), corresponds to a concentration of only 2.9 ppG. For example, fish total only some 0.47 ppG of all matter in the ocean, that is, 0.47 grams of "fish" in each 1000 m³ of water. Compare this to uranium, one of the rarest elements on the planet, which has a concentration of 3.3 ppG, or 7 times more than the fish in the sea, equal to 4.5×10^{12} kg of uranium in ocean water.

It is obvious that not just the mean concentration of specific constituents is of importance. The case of fish versus uranium is a good example. While accepting the mean concentration of 0.47 ppG fish of various types, at different times and in different areas (particularly near the surface), local concentrations may be much higher. Uranium, however, is more or less homogeneous throughout the ocean at all times and at all depths.

Drinking water is different from ocean water and closer to rainfall and the fresh water of rivers. The difference is most significant in the content of soluble

5.3 The Hydrosphere – A Crucial Factor in the Existence of the Biosphere

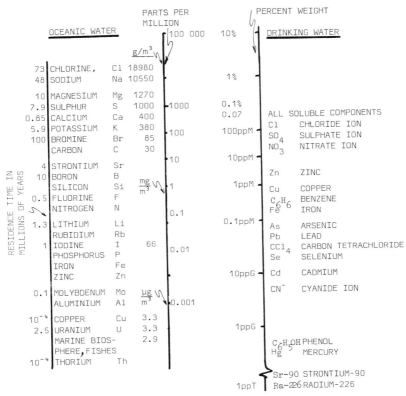

Figure 5.11. The chemical compositions of the sea and drinking water.

substances, ocean water containing about 3.5 weight-percent of solubles, mainly chlorides and sulphates of sodium and magnesium, whereas good drinking water has not more than 0.07 weight-percent, that is, 50 times less. In real life much greater difficulties are experienced with the man-made substances we find in our drinking water. Man-made carbon tetrachloride in drinking water should not exceed 0.1 ppM and phenol even less than 1 ppG, that is, less than one gram of phenol in 1000 m³ of water.

An even more severe limitation is found in the case of radioactive substances. In natural, non-contaminated drinking water, the concentration of the naturally occurring nuclide radium-226 must not exceed approx. 37,000 Bequerels/m³. In more direct units, the amounts of radium-226 must be small exceed 37,000/second in 1000 m³ of water, which corresponds to ~1 μg per m³, that is, approx. 1 ppT.

5.3.3 Man's demand for water is gigantic

A man drinks approx. 2 kg of water per day, or 0.7 m³ per year. Adding his needs for washing, cooking, agriculture, and industry as well as for cooling in

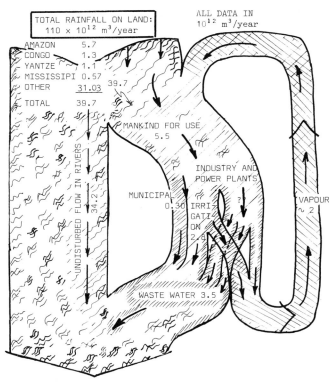

Fig. 5.12. Water balance on a global scale. Prognosis for the year 2000 (according to FAO).

power stations, the average per capita annual demand is some 1500–1800 m^3 (in the USA probably 3000 m^3).

Taking America as an example, we can assume that man uses ~ 3000 m^3/year. If the mean rainfall equals a 0.7-m layer of water and allowing 70 % evaporation, an American needs a surface area of 14,000 m^2 to supply his need for water. The total area available to him at present is 44,000 m^2.

It is even more interesting to look at our case here on a global scale. Some scientists claim that in the year 2000 the population of some 6.1 billion people will use 6×10^{12} m^3 of water annually, that is, 1000 m^3 per capita (Figure 5.12).

5.3.4 Drinking water, where purity counts

The amount of water directly consumed by man forms only a small part of the total water demand and equals roughly 1 m^3 per annum per capita, or about 0.05 % of the total. The whole problem, however, is one of quality (Figure 5.11).

The purity of drinking water is determined by two parameters, the bacteriological purity and the chemical purity. In this book the first problem is not dealt with, but we shall consider some aspects of the chemical impurities of drinking water. It is perhaps worth pointing out beforehand that the problem of the cleanliness of drinking water and of the air we breathe is not a new problem, but one that is as old as mankind itself. The cave dweller suffered from very bad air in his cave, particularly carbon monoxide and smoke from the fire. Drinking water was polluted by his bodily wastes, giving rise to bacteriological activity. Even the citizen of Rome suffered from tin and lead poisoning from his drinking vessels. Pollution is nothing new.

5.3.5 The erosion of the planetary surface

At the present time, a rather large amount of terrestrial soil is transported by the rivers into the oceans. Each year, the Yellow River in North China carries $\sim 500 \times 10^9$ kg of soil, the Yangtze approx. 200×10^9 kg, and the same amount by the Mississipi. How large is the worldwide total?

The best estimate is that at the present time, all the rivers on the Earth dump approx. 5×10^{12} kg of soil into the ocean annually. The global annual denudation during the Tertiary Period has been estimated at 5×10^{12} kg, which approximately equals the present rate.

The total amount of sedimentary rocks at the present time has been estimated as equal to 3.2×10^{21} kg, which corresponds to an average world layer of approximately $\sim 6 \times 10^6$ kg per square metre, that is, a thickness of 2.2 km. The present erosion rate of continents, estimated by direct measurements, results in a loss of approx. 0.03 mm/year (for the United States ~ 0.06 mm/year).

5.4 The Solid Earth, the Lithosphere

5.4.1 The main components of the Earth's crust

The matter in solid form in the Earth's lithosphere contains 99.977 % of the total terrestrial matter (Figure 5.1). The crustal envelope with a thickness of 20 km contains 2.4×10^{22} kg, that is, 0.4 % of the total lithosphere. About half of this gigantic amount of material in the Earth's crust is oxygen in the ratio of 63 atoms of oxygen to 100 atoms of all elements in the crust. All of the other 90 elements are represented by the remaining 37 atoms. On a volume basis, however, the ratio is even greater. Compared to its partners silicon and other metals, oxygen takes up 93.7 percent leaving only 6.3 percent for the rest. Really the Earth's crust should be called the oxygen sphere (Figure 5.13).

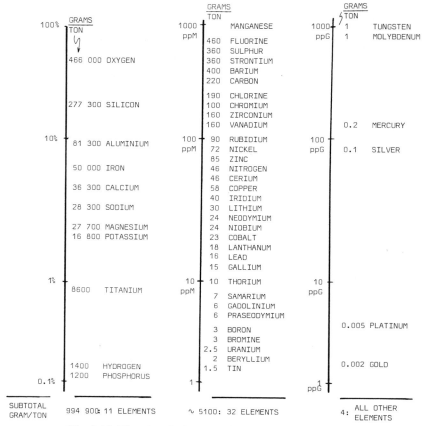

Fig. 5.13. The chemical composition of the Earth's crust.

The crust is made up of some 2000 different minerals. Only some 80 of these are of major importance, making up about 95 percent of the total. The remaining 5 percent is made up by the other ~ 1900 minerals.

5.4.2 The Earth's crust, the main source of materials for our civilisation

The size of the hydrosphere has already been discussed together with its role as a source of water for mankind. All other materials are, relative to water, used by Man in significantly smaller amounts, about three orders of magnitude, or one thousand times, less. Nevertheless, the absolute amounts are large.

Omitting the atmosphere as a source of oxygen for sustaining life and for burning fossil fuels, together with the nitrogen from the atmosphere fixed to give fertilizers, the lithosphere becomes the main source of the remaining materials, in which the biosphere plays a major role (Figures 5.14 and 5.15).

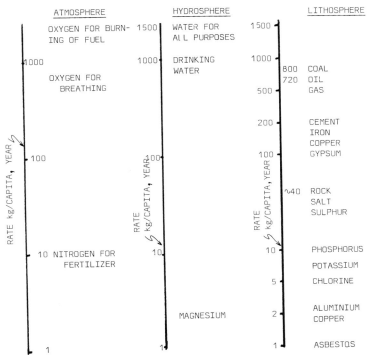

Figure 5.14. The annual material flux from the atmosphere, hydrosphere, and lithosphere per capita.

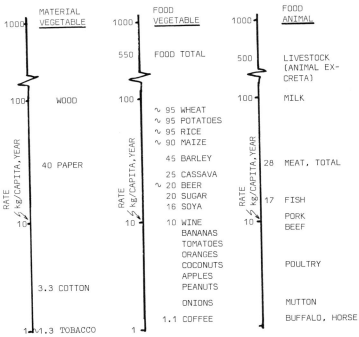

Figure 5.15. The annual material flow from the biosphere per capita is also very large.

The material flux should be divided into three classes:

- principally non-recycled material, such as fuel and food;
- principally recycled materials, such as structural components, metals, and paper;
- practically non-recycled materials, such as cement and wood.

Of course, this division can be questioned on several grounds; for example, oil products are used not only as fuel but also as raw material for the production of, for example, synthetic fibres.

Problems of fuel are discussed in Chapter 6, food in Chapter 7. In the next section we deal with the other materials used in large amounts, beginning with the metallic elements.

5.4.3 Metals "prepared" by Nature, the most widely used

Our civilisation is tightly bound up with the use of metals. Our technology is first and foremost based on machines, apparatus, vehicles, buildings, bridges, tubes, wires, etc., made of metals.

The metallic elements must fulfill particular criteria. The formulation of these has been built up not in a rational way, but rather haphazardly. Some of the criteria for selecting metals are, however, self-evident, that is, the need for:

a) special mechanical, electrical, chemical, and thermal properties;
b) chemical stability in the atmosphere;
c) ease of recovery and manufacture;
d) abundance of the high-grade ores;
e) economic cost of preparation.

With this wide spectrum of selection criteria, it will be surprising to note that the present civilisation's use of metals is more or less connected with the properties and abundance of cosmic matter. The following remarks will make the situation clearer.

In spite of the very complex history of cosmic matter during the accumulation and evolution of the Solar System and the loss of more than 99 % of the volatile components, the abundance of most metals in the Earth's crust is very similar to the cosmic abundance, as a comparison of Figures 5.13 and 2.4 shows.

The next point seems perhaps less significant, but it is a fact that the production rate of these metals in our civilisation is more or less proportional to their terrestrial abundance. We are using, in the first instance, those metals provided in the greatest abundance (Figure 5.16).

Thirdly, the cost of metals being produced is also roughly in proportion to their terrestrial "rarity", that is, the less abundant materials are more expensive (Figure 5.17).

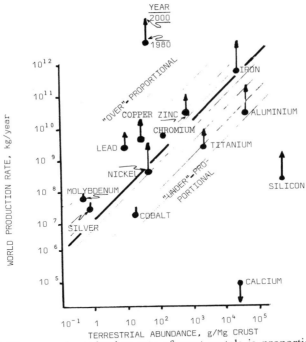

Figure 5.16. The annual production rate of most metals is proportional to their terrestrial abundance.

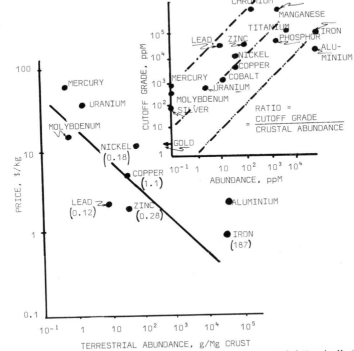

Figure 5.17. The price of metals is proportional to the terrestrial "rarity". (Probable resources in 10^{12} kg are shown in parentheses.)

5.5 Ordered Matter and Entropy

5.5.1 Concentration means increase of order and decrease of entropy

Many technological processes in the production of materials begin with the concentration of matter.

Most elements important to our civilisation are present on Earth in the form of chemical compounds that are more or less randomly distributed in the crust. The technology at this stage is concerned with two concentration processes:

a) concentration of the chemical compound which carries the metal in question, which is the separation of the compound from inert material (other minerals in the rock);
b) concentration of the metal itself, which is the separation from the additional components of the chemical compound (e.g., oxygen, sulphur, and carbonate).

From the point of view of physics, both of these processes are very similar; they increase the concentration of the element by separating it from the associated elements. This is clearly increasing the order of the material and decreasing disorder; in other words, concentration decreases entropy.

We have seen that the normal direction of events in the Universe is a move towards increase of entropy, that is, decrease of order, and any process going against this must be powered by a corresponding parallel process which provides the necessary energy by moving in the direction of increase of entropy. This second process is the flow of highly ordered energy from a so-called energy source to an "energy sink", whereby the energy is degraded and transferred into a disordered state with a high entropy content.

Thus, to obtain the material required from natural sources a source of free energy is required. This, too, comes from Nature and is not "gratis". Here we see how the pricing of various materials is governed partly by the amount of energy required to extract and concentrate them.

The unit energy required to produce a unit amount of given material ranges from 40 MJ (megajoules) per kg in the case of lead, up to more than ten times greater, that is, 500 MJ/kg, for production of titanium. The most abundant element in our technology, iron, requires 60 MJ/kg (Figure 5.18).

Recycling of metallic wastes, that is, the recovery of used metals, needs significantly less free energy by about a factor of 10 than that required for the original extraction. Similarly, the production of other nonmetallic substances such as primary chemicals requires large amounts of energy. For example, the synthesis of ammonia from nitrogen and water uses ~ 100 MJ/kg, and even the production of cement requires 8 MJ/kg.

5.5 Ordered Matter and Entropy

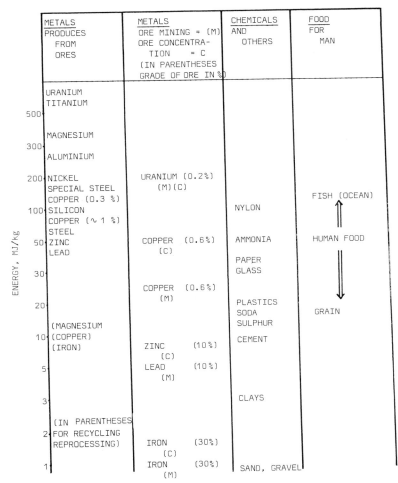

Figure 5.18. The free energy needed to produce some materials.

5.5.2 Impact of substances in very small amounts: Poisons

This planet in its natural state includes a rather small number of chemical compounds. As we have seen, the number of minerals is approx. 2000. The number of biogenic compounds can probably be estimated as less than 10,000 (very roughly!). But at present the number of man-made substances increases by approx. 1000 each day. The best source of information, *Chemical Abstracts*, gives some data on the man-made substances, which now number approx. 4×10^6 compounds. Most of them are made or separated only on a laboratory scale. But at least 63,000 compounds are chemicals in everyday use.

The most dramatic situation can be seen in the production and use of a very dangerous class of chemicals – the pesticides. The present annual production of pesticides equals $\sim 5 \times 10^9$ kg, more than 1 kg per capita per year. This shows a tendency towards a further significant increase. These and other chemicals seem to be among the most dangerous products of global significance. In the USA in 1980 industry generated over 50×10^9 kg of hazardous chemical wastes and disposed of most in the environment.

5.5.3 Material dissipation and waste formation increases entropy

In the flow of materials there are partial processes which proceed spontaneously without the need for any input of free energy from the outside. These are only possible when an increase of entropy occurs as a result. These processes include dissipation, decrease of concentration, destruction of internal order, destruction of macroscopic shape, weak chemical reactions with the atmosphere (corrosion), strong processes of oxidation (fire), and many others that produce a great deal of waste.

Not only these spontaneous reactions produce waste. Man's technology has been established for centuries to produce selected materials and components. Inextricably, by-products are also formed, which are mostly classified as wastes. Clearly the efficiency of our technology is rather low. Further, large amounts of waste arise from the extraction of energy from fuel and food.

Much of this waste production is an inherent part of our life and technology. However, much can be done to recycle waste and transform it back into useful materials, with the exception of fuels themselves. The magnitude of the problem can be seen from Figure 5.19, in which the amount of material used by the average American during his 70-year life is shown. Here we see that one man with a body weight of, say, 70 kg "uses" in total about 70,000 Mg or one million times his body weight in material (including material moved in mines, urban activity, agriculture, water consumption, and irrigation) during his life. Each hour the weight of matter "used" is greater than that of his body.

The question is, then, for how long can we live in a civilisation which has a throughput of material on this scale, producing at the end so much waste or waste-like matter?

In theory, the solution is simple, but difficult to realise. In principle, all these waste products can be recycled in a fully integrated system together with the natural flow of material. The price to be paid is the use of free energy. It seems a reasonable price which mankind in the future will be able to meet.

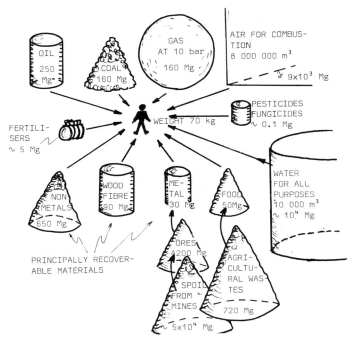

Figure 5.19. The use of food, fuel, water, metals, and other material by the average American during his 70-year life.

5.6 What Are the Conclusions for Mankind's Future Development?

This chapter contained material on:

- the indestructibility of terrestrial matter;
- the origin and composition of cycles in the atmosphere;
- the origin and composition of cycles in the hydrosphere;
- the influence of man on the material cycles in the lithosphere;
- the relationships between the flow of free energy and the properties of matter.

For the theme of this book – the future of mankind – the following conclusions can be drawn:

- All stable and quasi-stable chemical elements existing in Nature are available at the Earth's surface and permit "unlimited manipulation".
- The material cycles, although largely due to natural forces, can be controlled by Man.

- The present trend of increasing use and manipulation of natural resources must be better controlled and guided in the future.
- A well-developed use of material cycles can allow further technological evolution of civilisation.
- A sophisticated materials cycle requires the availability of a large amount of free energy, but the amount of energy presently known appears to be adequate.
- There is no danger of exhaustion of material resources in the future.

The ultimate future of mankind depends on Man himself.

CHAPTER 6

The Flow of Energy on the Earth

> In the huge manufactory of natural processes, the principle of entropy occupies the position of manager, for it dictates the manner and method of the whole business, whilst the priciple of energy does the bookkeeping, balancing credits and debits.
>
> R. Emden
> (1862–1940)

> What we call an energy crisis really ought to be called an available energy or even free-energy crisis.

> If we ask why Man has, ever since the beginning, chosen to use more energy, it is probably because he values time more than he does energy. Remember that thermodynamics demand immortality of us if we are to conduct our lives with minimum expenditure of energy. Reversible processes require infinite time or infinite capital energy ... by expending energy, we can save time.
>
> A. Weinberg
> (1915–   )

6.1 The Sources of Free Energy on the Earth

6.1.1 The quality of energy: The ordered and disordered forms

The history of the Universe and its component parts – living beings, stars, molecules, and atomic nuclei – is the history of the intimate coupling of particles and free energy.

The objects in the Universe differ from one another not only in terms of mass or chemical composition but also in the level of their organisation. The higher the level of organisation, the higher the order and the lower the likelihood of the spontaneous formation of such an object.

The persistence of these highly ordered objects is maintained by the influx of free energy. Each ordered system deprived of this flux of free energy loses its internal order and decays to a disordered state.

It is clear right from the start that the Universe consists of more or less ordered objects from the clusters of galaxies to the living, thinking being. All these are supported by a flux of ordered energy from various sources flowing through the process of "ordering", ending up in the "sink" of disordered energy.

6.1.2 The elementary forms of energy

It has been shown that four elementary forces, or interactions, can be classified as:

	Strength
a) strong (also called nuclear)	~ 0.3
b) electromagnetic	0.073
c) weak	$\sim 10^{-12}$
d) gravitational	$\sim 10^{-40}$

However, the sources of energy as described in Chapter 1 are reduced to two types – those due to nuclear and those due to gravitational interactions. These are coupled together in cosmic objects, for example, in the stars. It is these two primary sources of energy which are important to the planet Earth.

Before going into detail, it is useful to classify energy into two further classes, the continuous supply or flux of energy, and the "stored" energy.

Of the first type, the continuous flux of primary energy of importance to the Earth, two forms exist:

- the nuclear processes (slow hydrogen burning) in the Sun, transported to the Earth by electromagnetic radiation ranging from the ultraviolet through visible to the infrared, and
- the gravitational interaction of terrestrial oceans with the Moon and the Sun.

The second type of energy, existing as stored energy, is also of two kinds:

- the nuclear energy of fusion of light elements, that is, hydrogen (and deuterium), lithium, beryllium, and boron, and
- the nuclear energy of fission of the heavy elements uranium-235, but also, in principle, uranium-238 and thorium-232. Included here is a large part of the geothermal heat flux, which has the same origin.

The fusion energy carriers hydrogen and deuterium are examples of products formed in the very early Universe, in the first seconds of the Big Bang, approx. 15 gigayears (15×10^9 years) ago. The fission energy carriers

6.1 The Sources of Free Energy on the Earth

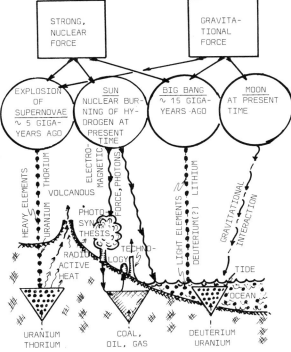

Figure 6.1. Primary sources of free energy.

uranium and thorium were produced much later in the outer envelope of supernovae, approx. 5 gigayears ago.

All other forms of energy, whether continuous or stored, are derivatives of one or another of the primary energy sources. Figure 6.1 shows the energy forms of importance for the Earth and for mankind in a simplified manner.

6.1.3. How large is the flux of energy?

For a better understanding of the problem of energy and its role in Nature, it is useful to consider the magnitude of the energy flux in the Universe.

In spite of incomplete information it can be estimated that the total flux in the Universe from source to sink is approx. 10^{48} W. This value is obtained by assuming that the Sun is a typical example of the average component of cosmic matter. The Sun produces 4×10^{26} W.

How large is the total flux per atom in the Universe? Taking the above cosmic figure, the mean flux equals:

$$\frac{(10^{48} \text{ W/Universe}) \times (6.2 \times 10^{18} \text{ eV}/(\text{W} \cdot \text{s}))}{\sim 10^{79} \text{ atoms/Universe}} = \frac{\sim 10^{-12} \text{ eV/s}}{\text{atoms}}.$$

For the 15 billion years of the Universe's existence the total amount of energy emitted by a mean atom equals:

$$\left(6 \times 10^{-13} \frac{\text{eV/s}}{\text{atom}}\right) \times \left(3.15 \times 10^7 \frac{\text{s}}{\text{year}}\right) \times (15 \times 10^9 \text{ years}) = 3 \times 10^5 \text{ eV/atom}.$$

Since the binding energy per nucleon in an atomic nucleus equals some mega-electron volts, the result giving 300 keV/atom in the whole history of the Universe seems to be reasonable under the condition that the flux intensity during the whole period was about the level observed today.

The energy flux of the Sun at the present time equals 3.72×10^{26} W, as has been mentioned, or 3.72×10^{14} TW. The Earth receives only 4.7×10^{-10} part of this flux, that is, 175,000 TW. All other energy flows on the Earth are less than this. The most important in terms of power are (Figure 6.2):

- total energy flow in the evaporation of water,   44,000 TW
- total net energy flow converted into chemical energy in the biosphere,   92 TW
- total energy flow in the technological activity of mankind,   10.5 TW
- energy flow in the total metabolism of mankind.   0.5 TW

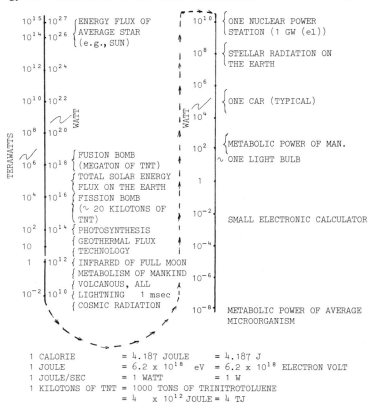

Figure 6.2. The scale for energy flows (power).

6.2 The Energy Sources on the Earth

6.2.1 Solar energy – The most important source

A rather complicated technique must be used for the direct measurement of the energy flux from the Sun onto the outer layer of the terrestrial atmosphere.

The best data give a value of 1367 W/m². This value, known as the "solar constant", is known to an accuracy of 0.5%. It should be said that the true value varies from 1413 W/m² at the perihelion (the closest approach to the Sun) to 1321 W/m² in at the aphelion of the Earth's orbit. This is equal to a 3.3% change over the year compared to the average value.

Since the cross-sectional area of the Earth is given by

$$\pi R^2 = \pi (6.378 \times 10^6 \text{ m})^2 = 1.278 \times 10^{14} \text{ m}^2 \quad (R = \text{radius of Earth}),$$

to total energy flux from the Sun equals

$$1367 \text{ W/m}^2 \times (1.278 \times 10^{14} \text{ m}^2) \cong 1.75 \times 10^{17} \text{ W} \cong 175{,}000 \text{ TW}.$$

This is only a first approximation. Due to reflection from the outer layer of the atmosphere, clouds, dust, and the Earth's surface, a good part of the solar energy is reflected back into cosmic space. The coefficient of the reflectivity, called the albedo, for the Earth averages approx. 0.310 (see Figure 6.5). This means that for the total energy flux of 175,000 TW only 69% – that is, 120,500 TW of solar energy flux – can be absorbed by the atmosphere, hydrosphere, and surface. The simplified balance of solar energy flow on the Earth is given in Figure 6.3.

The real balance for the Earth is much more complicated due to the following factors:

– There are systems which can store thermal energy in the hydrosphere and atmosphere, and these can mutually exchange some of the absorbed solar energy.
– A significant part of the solar energy is converted to other forms (e.g., mechanical energy) in the moving atmosphere (winds), energy in the movement of the hydrosphere (waves), the chemical energy of photosynthesis, and heat which evaporates water from the biosphere.
– There are some small, but not negligible, sources of energy which are completely independent of solar energy (see Section 6.4.2).

All these factors result in a rather complex and still imperfectly understood system of energy flows which has existed in stable form for millions of years. However, recently these have become more and more influenced by Man's own technology.

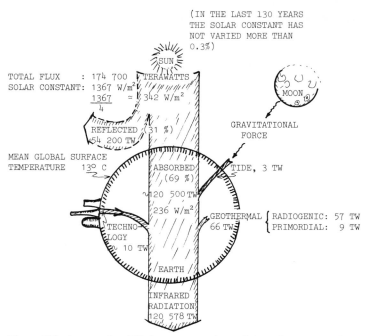

Figure 6.3. A very simplified representation of the solar energy flux.

6.2.2 Spectrum and albedo of solar light

Reflection and absorption result in changes in the spectrum of solar energy at the Earth's surface. The undisturbed extraterrestrial solar flux, with an intensity of 1367 W/m^2, has a spectrum as shown in Figure 6.4. The total solar flux on the Earth's surface equals $1367/4 = 341.7$ W/m^2.

The rest of this energy flux reaching the Earth's surface is not only reduced by a factor of 2, but is also significantly changed, as Figure 6.4 shows. The most dramatic and most important change is in the reduction of the ultraviolet part of the spectrum in the range near 0.35 µm, that is, 350 nm. Diffuse daylight, which is very weak, has also lost much of the infrared part of the spectrum.

The night side of the Earth emits infrared radiation equivalent to a blackbody radiation of 255 K, that is, $-18\,°C$. Figure 6.4 shows this radiation as a spectrum having an intensity of 240 W/m^2, which gives a total terrestrial albedo of:

$$\frac{341.7 - 240}{341.7} = \frac{\sim 106}{341.7} = 0.31.$$

Of this solar flux of 240 W/m^2, only ~ 168 W/m^2 reaches the Earth's surface. The rest is absorbed by the atmosphere.

In spite of the fact that the average temperature of the Earth as seen from space is 255 K ($-18\,°C$), the average surface temperature is 288 K, that is,

6.2 The Energy Sources on the Earth

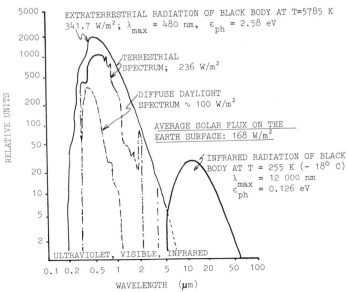

Figure 6.4. The solar energy spectrum versus the night emission of the infrared. ε_{ph} = energy of photon for maximum of spectrum.

15 °C. This can be partly explained by the greenhouse effect, caused by atmospheric water vapour and carbon dioxide. The greenhouse effect would be enhanced by a higher CO_2 concentration.

The value of the total albedo of the Earth is given in Figure 6.5. Unfortunately, a lot of uncertainties and discrepancies influence the exact calculation of the albedo.

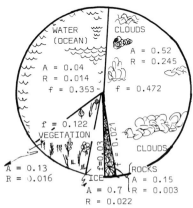

Figure 6.5. The albedo of the present Earth. f = fraction of earth's surface; A = albedo; R = reflectivity; $R = f \times A$; R = (earth) = 0.310; according to Budyko; R (earth) = 0.330.

The most important component affecting the Earth's albedo is the clouds. The proportions of the different types of clouds – cirrus, altostratus, altocumulus, and cumulus – are not sufficiently well known. Also, the albedo of the snow-covered surface is not well known and seems, in the years 1968–1979, to have increased from 8.2% to 10.1%. In the Northern Hemisphere in 1980 this value decreased to 8.9%.

Man-made changes of albedo are also important. The following table shows these values.

Process		Albedo before	Albedo after	Percentage of Earth's surface
Savanna	→ Desert	0.16	0.35	1.8%
Forest	→ Grassland	0.12	0.15	1.6%
Forest, Field	→ City	0.52	0.15	0.2%

Man has changed 17% of the continent's surface, that is, ~5% of the Earth's surface. This probably changes the global albedo from 0.305 to the present value of 0.310, corresponding to a decrease in solar flux of approx. 860 TW. The result is a global cooling of ~1 K.

6.3 Solar Energy and Climate

6.3.1 The solar energy flux is not constant

The Sun is a Main Sequence star (see Chapter 3). These stars evolve rather slowly over a period of gigayears. A star having a mass of 10^{30} kg (solar mass: 1.98×10^{30} kg) evolves over 10 gigayears. Earlier it had been thought that the energy flux reaching the Earth was constant over long periods. However, over the 5-gigayear history of the Earth, it seems that the Sun has continued to evolve and the solar flux reaching the Earth has increased by 20% over that period.

It is now known that the flux cannot be constant, the reasons being summarised in Table 6.1. The most important parameters are shown in Figure 6.6 according to the so called "Milankovitch hypothesis".

The change in the solar constant is probably responsible for the long-term changes in the terrestrial climate. Of the greatest significance is the fact that the solar energy flux on the Earth's surface is unevenly distributed. The mean value can be estimated in the following manner.

6.3 Solar Energy and Climate

Table 6.1. Parameters Affecting the Absorption of Solar Energy

Cause	Parameter	Details	Duration or period	Consequences
Sun	As a star	Solar spots, short term Solar spots, long term Solar flares Star evolution	11 years hundred years ?? billion years	Colder climate? Thunderstorms, rains? Solar constant[a] more heavy metals on solar surface, resulting in more ultraviolet radiation
	As a star in the galaxy oscillating perpendicular to galactic plane	Clouds in the galactic arms	28 million years	
Companion star of Sun Remark: Hypotheses without observational facts	Dwarf star (0.01 solar mass) maximum distance to Sun approx. 2.5 light-years	When the companion passes close to the "Comet cloud" it causes a shower of comets in the vicinity of Earth	~28 million years	The impact of the comets with the Earth threw up enough dust to darken the Earth's atmosphere for a least six months inhibiting photosynthesis
Sun–Earth mutual position	Cosmic space	Large comets (10 km diametre)	28 million years	Dust, clouds
	"Milankovitch hypothesis"	Eccentricity of orbit Obliquity of axis Precession of perihelion	100,000 years 41,000 years 23,000 years	See Figure 6.6

Table 6.1. (Continued)

Cause	Parameter	Details	Duration or period	Consequences
Earth	Day/night rotation	Present rate of Earth rotation: 24hrs	After some billion years increased to 30hrs	
	Atmosphere	Ozone	Years	Cutting the ultra-violet radiation
		Carbon dioxide	Hundred years	Greenhouse effect
		Dust (volcanoes, desert)	Years	Albedo changes
		Clouds	Very short	
	Biosphere	Vegetation	Thousand years	Albedo changes
	Hydrosphere	Ice sheet (cryosphere)	Thousand years	Glacial periods
	Continents	Movement of continents (pole wandering)	Million years	Changes in currents (see Figure 6.8)
		Growth of mountains, Sea level changes	Million years	Changes in winds
	Technosphere	Dust, land use, carbon dioxide (fossil, fuel)	Decades	Albedo changes
		Spray propellant Heat pollution	Decades	Greenhouse effect Ozone destruction

Solar constant	Percent	Consequences
Increase	2%	Total melting of ice sheet
Decrease	4%	Extension of snow and ice cover are over the entire globe

[a] Remark:

6.3 Solar Energy and Climate

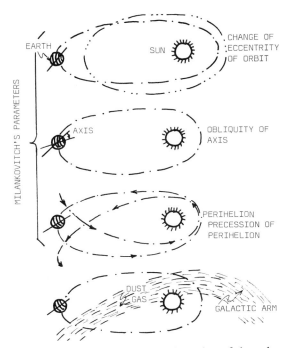

Figure 6.6. The parameters influencing the value of the solar constant.

$$\left(\frac{\text{solar constant}}{\text{ratio of surface/cross section area}}\right) \times \left(\frac{\text{solar flux on surface}}{\text{solar flux on the atmosphere}}\right) =$$

$$\left(\frac{1367}{4}\right) \times (0.49) = \sim 168 \text{ W/m}^2.$$

The real average solar flux on the surface changes from less then 50 W/m² for the polar regions to more than 260 W/m² in the equatorial region and subtropical deserts.

The regions having the highest population density have a solar flux of approx. 100–160 W/m². This is the limiting factor for agricultural activity (see Chapter 7) and for the future direct use of solar energy (Figure 6.14).

6.3.2 Solar energy is transformed into numerous forms and types of energy

The highest state of energy into which the solar flux is transformed is, without doubt, the 92 TW of net power transformed into chemical processes in the biosphere. Also of rather high quality is the mechanical movement of air, water, winds, and waves. They total 3300 TW.

The largest part, 44,000 TW, is used to evaporate 1.6×10^{10} kg of water per second. A small but not negligible part of that energy is transformed into electrical energy (lightning). This form of free energy may have been rather underestimated up to now.

All these forms of energy and matter result in the phenomenon called "climate" (Table 6.2). Many parameters influence the climate machine and make it complex and difficult to understand or predict (Figure 6.7).

Table 6.2. Heat Contents of Climate Machine

Storage of thermal energy	Mass of storage	Specific heat (J/kg · K) or heat of melting	Temperature difference (arbitrary values)	Heat contents	
					in time units of solar absorbed energy flux (years)
	(kg)	(J/kg)	(K)	(TJ)	
Ocean	1.4×10^{21}	4187	5	2.9×10^{13}	~7.6
Ice (glaciers) $\sim 15 \times 10^6$ km^2 thickness: 3 km	4.6×10^{19}	heat of melting 344,000	0 (melting)	1.5×10^{13}	~4.0
Air	5.1×10^{18}	~1,000	15	7×10^{10}	~1

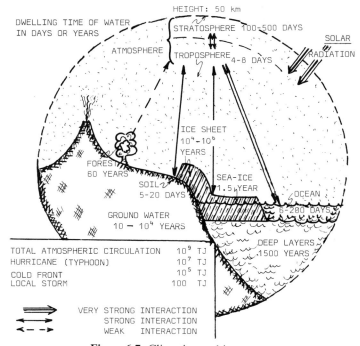

Figure 6.7. Climatic machine.

6.3 Solar Energy and Climate

6.3.3 The past and future of the terrestrial climate

The climate is not only an unpredictable thermodynamic machine but was also rather capricious in the past. What is likely to happen in the near and the distant future?

The present climate at latitude 10° North has a mean temperature of 3° Celsius, which is rather high. In the past half-million years the climate was hotter than now for only 8% of the time. Each of these cold periods lasted approx. 50,000 years (Figures 6.8 and 6.9). The glacial periods have significant impact on the total Earth's surface (Table 6.3).

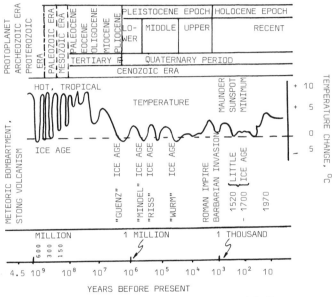

Figure 6.8. The evolution of the terrestrial climate.

Table 6.3. Impact of Glaciations

Parameter	Unit	Present time	Maximum glaciation	Minimum glaciation
Continental ice sheet	million km³	25	75	0
Land surface covered by ice	percent	10	30	0
Surface of continents open to biosphere	million km²	117	90	~128
Ocean level	m	0 present level	−150	+55
Albedo of Earth	ratio	0.310	>0.31	<0.31

The longest period of global ice-free climate lasted from 65 to 140 million years ago, during the Cretaceous Period.

This may have been the only ice-free period in the Earth's history. At the end of the Cretaceous Period, over a very short time, dramatic changes occurred and more than half of the species on Earth disappeared. This time is known as the "Great Extinction" (see also Figures 7.8 and 7.9).

Over the last thousand years, the coldest period occurred in the 17th century. From this time up to 1940 the mean temperatures in the northern latitudes increased by about 1.7 K, but the annual fluctuations increased up to ± 2.2 K. In the more distant past the climate changed significantly. Nevertheless, both of the extremes have been avoided – a cold, ice-covered planet or a very hot, waterless planet (similar to Venus) – by very narrow margins.

What of the future? The best prognosis claims that the probability of an ice age occurring in the next 100 years is 0.002. Most probable for the next 100 years is a warming of the terrestrial climate.

There is some suspicion that the climasphere has not just one stable state, even when all significant external influences do not change. It could be that the system has two or even more metastable states. If so, a change from one to another might be triggered by relatively minor upsets. These changes could be influenced by human activity. The climate itself has an overwhelming effect on most activities on this planet, including the oceanic and atmospheric activity, erosion of the Earth's soil, the activity of the biosphere, and, last but not least, human civilisation. A better knowledge of climatic mechanisms is of the highest importance for the future (see Chapter 8).

6.3.4 The local climate depends on continental drift

It is extremely trivial to state that local climate near the poles is different from that near the equator; therefore, the most important parameter influencing the local climate is the geographical latitude. It is probably not so trivial, and is now a well-established fact, that the continents, Asia, Europe, America (South and North), Africa, Australia, Antarctica, as well as India and Greenland, have had a rather adventurous history in the past 500 million years (Figure 6.9).

In the Cambrian period – that is ~ 500 million years ago – Europe, North America, Siberia, Antarctica, and Greenland were below the equator. The intensive increase of primitive plants in these past tropical regions can be seen today in the coal layers.

The self-evident question is the following: What is the cause of such a significant and dramatic drift of continents? What forces, what energy sources, are responsible for this phenomenon?

The answer is neither easy nor obvious, but currently there are some more or less well-founded hypotheses for the elucidation of this drift of continents, that is, a drift of a mass of approx. 10^{22} kg over more than 500 million years.

6.3 Solar Energy and Climate

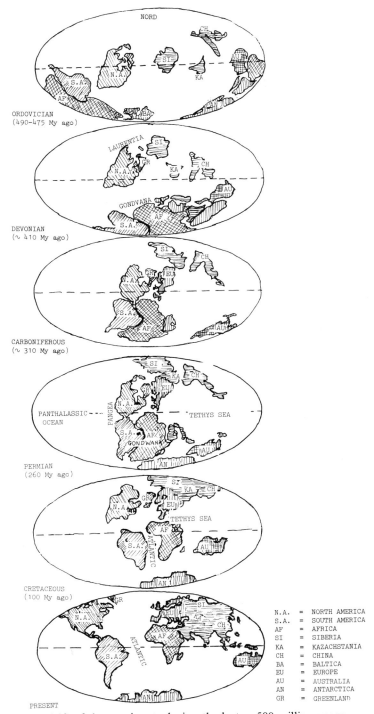

Figure 6.9. The drift of the continents during the last ∼ 500 million years.

There is only one significant energy source able to supply the required force:

- the geothermal heat flux, currently 66 TW, and in the distant past probably even 4–7 times greater.

6.4 Non-solar Terrestrial Energy Sources

6.4.1 Other non-solar flows of energy play a small but not insignificant role

As shown in Figure 6.1, the Earth has access to energy not only from the Sun but also from other independent sources. (Table 6.4 shows, without explanation, the role of these non-solar sources.) However, nothing can be further from the truth than to claim that this minute fraction of energy from other sources can be neglected.

First, consider geothermal energy, which is continuously produced by the spontaneous decay of the four long-lived nuclides, potassium-40, thorium-232, uranium-235, and uranium-238, and their corresponding decay products. The total global flux from this source is 66 TW, that is, only 0.11 W/m^2. Probably 9 TW results from the cooling of the Earth's core (primordial heat). All these values are uncertain.

Despite this small value, the influence of this source on evolution on this planet is extremely important. The following phenomena are direct products of this geothermal heat flow:

- continental drift and, for example, the formation of the Atlantic Ocean more than 100 million years ago (see Figure 6.9). The present rate of ocean bottom formation is $\sim$3 km^3/year, which corresponds to an energy flux of $\sim$0.2 TW;
- volcanic activity, explosions, dust production, and the resulting changes in the Earth's albedo and the transmissivity of the atmosphere: there are

Table 6.4. Non-solar Energy Sources

Energy source on the Earth	Flux (TW)	Relative to solar energy flux
Solar energy absorbed by the Earth	120,500	1000.00
Geothermal energy (radioactive decay plus primordial heat)	66	0.54
Tidal energy (gravitational forces)	3	0.025
Technology (fossil fuel and nuclear energy, man-made sources)	$\sim$10	$\sim$0.08
Total for night emission (see Fig. 6.3)	120,579	1000.64

6.4 Non-solar Terrestrial Energy Sources

10,000 "active" volcanoes although only 500 have actually erupted in historic times. The present lava production of $\sim 1 \text{ km}^3$/year corresponds to an energy flux of ~ 0.16 TW; total volcanic heat: 0.8 TW;
- earthquakes (average rate corresponds to 1 TW);
- geysers and thermal springs.

For these reasons, and due to the poor thermal conductivity of the Earth's crust, geothermal sources have played an overwhelming role in the terrestrial evolution. The maximum known earthquake, with a magnitude of 8.6 on the Richter scale, corresponds to an energy release of approx. 500 PJ (petajoules), equivalent to the total solar energy absorbed during a four-second period on the whole Earth. The same energy could be released by the explosion of 25 thermonuclear devices of 5 megatons of TNT each.

Next, the tidal energy originating from the Moon and Sun's gravitational interaction is also small, but can have important local effects.

A rather significant parameter is the increasing amount of technologically produced energy, at present equal to ~ 10 TW, that is, 0.08 per thousand solar energy units.

6.4.2 The importance of the amount of stored energy

The most common characteristic of the energy sources discussed above is their relatively low energy density per unit area. Here the values of the specific energy flows per square meter are given (see also Section 6.10.6):

Energy	Specific flux density (W/m^2)
Solar on the Earth's surface (average)	~ 168
Geothermal	0.12
Gravitational	0.006

As an example it can be postulated that for a specific country the current energy demand is 8 kW per capita.

If solar energy could be used with an efficiency of, say, 10%, then the primary solar energy flux must have a value of 8000 W per capita to cover the above needs. With an average power density of 168 W/m^2 for solar energy flux,

$$\frac{8000 \text{ W/capita}}{168 \text{ W/m}^2} \text{ or } 48 \text{ m}^2$$

of very sophisticated solar collectors are needed to cover each person's energy needs.

Of course, at the present time, and even more so in the past, such a solar collector is far from economic. Mankind has, however, been given a very generous gift – this energy exists in a concentrated and stored form in fossil fuels and moving water in rivers. The amounts of the stored energy are given in Table 6.6.

6.5 How Much Energy Does Man Need?

6.5.1 Does man need energy at all?

Indirectly, what man ultimately needs is order. He needs order in manifold forms and in manifold ways. What is meant here by "order"? In the first place, man needs order in his own structure. It is known that each complex organism with a high order will decay spontaneously, producing an internal disorder.

Only the more or less continuous intake of free energy prevents the decay towards disorder and maintains the organism in its ordered state. For this purpose man needs a constant intake of energy – in the form of food – of about 10 MJ/day, that is, about 2400 kcal/day, equivalent to:

$$\frac{2400 \text{ kcal}}{\text{day}} = \frac{10 \text{ MJ}}{\text{day}} = \frac{10 \times 10^6 \text{ J}}{86{,}400 \text{ s/day}} = \sim 116 \text{ W}.$$

This amount of free energy permits man to maintain his internal level of order to support all necessary internal metabolic processes and to have energy for moving, working, thinking, and living (Figure 6.10).

Man also needs order in his immediate environment. He must maintain a reasonable temperature and humidity. He needs order for movement, for his

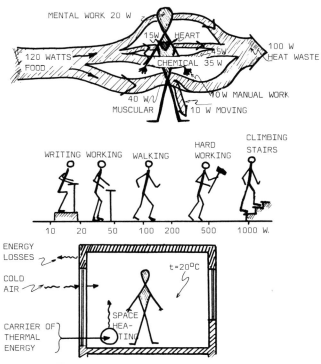

Figure 6.10. Energy for the human body.

6.5 How Much Energy Does Man Need?

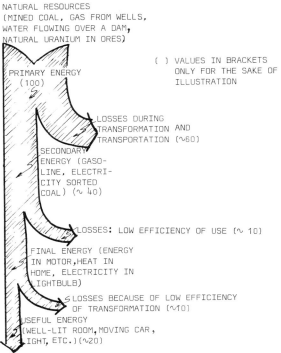

Figure 6.11. The flow of energy in civilisation.

Table 6.5. Energy Needs of Man in a Developed Society

		Watt/Capita
Food	Agriculture, Food industry (direct food distribution, input: 110 W), Cooking, etc.	800
Environmental heat	Space heating, Ventilation, Cooling, Conditioning.	1000
Housing	Construction of Houses, Hospitals, Schools, etc.; Maintenance of Houses, etc.	500
Transport	Construction of Vehicles, Construction of Roads, Fuel for Vehicles.	300
Culture, Science, School	Free-time activity, Information, Media, Cultural activity, Religion, Science	700
Social organisation, Military	State organisation, Police, Fire protection, Army.	700
Natural environment, Restitution	Recycling of Waste, Purification of Air, and Water, Protection from Weather, Natural parks.	1500
Subtotal	Energy for Direct Use	5700
Energy production, Distribution	Construction of Mines, Pipelines, Refineries, Power Stations, Transport of Fuel and Electricity, Processing of Fuel and Fuel Waste.	1300
Total	All needs	7000

social environment, and, last but not least, for the more distant environment in the global context. All these levels of order can only be achieved by the use of available free energy. Figure 6.11 shows the different forms of usable energy.

It is clear, therefore, that man needs energy to support his life and his development. The question now arises, how much is needed and what is necessary for an ordered "good life"? (Table 6.5)

6.5.2 The sources of energy are changeable

From the discussion in the previous sections it is clear that the development of Homo sapiens was connected with the kind and quality of the energy used. The evolution of these sources can be summarised as follows (Figure 6.12):

a) the energy of the human body;
b) wood, available on the Earth's surface and capable of self-sustained burning in the atmosphere. A wood-burning device is simple and the energy used to gather fuel, small;
c) draft animals;
d) hydro- and wind-power, in spite of very efficient generation, have never played a major role in any period of technological development, because of the sophisticated devices required and the geographical localisation of sources;

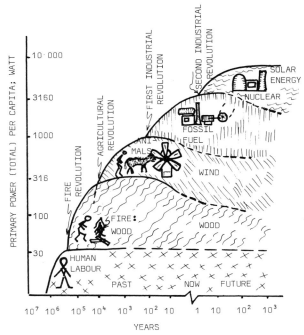

Figure 6.12. The energy fluxes and sources during man's history.

6.5 How Much Energy Does Man Need?

e) the discovery of the use of coal as a carrier of energy in the form of heat was of such significance that it triggered a new era – the Industrial Revolution. The more complex devices for burning coal and eventually transforming the resulting heat into steam and later into mechanical energy are very characteristic of this energy source. It is worthwhile pointing out that the existence of coal and oil on the Earth is a rather peculiar and unique result of the existence of the Carboniferous Period in the planet's history and the drift of continents (Figure 6.9). Without the Carboniferous Period the Industrial Revolution in the 18th century would not have been possible;

f) the need for still new energy sources, such as nuclear energy, demand very complex devices for releasing the energy from the atomic nuclei of heavy elements, and marks the present stage in the development of human technology.

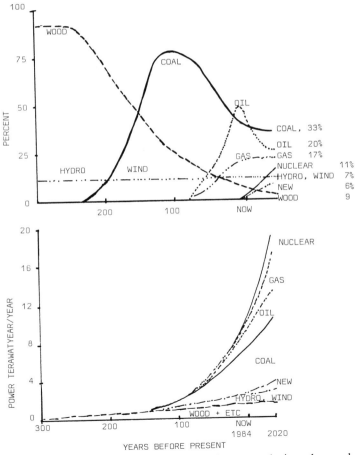

Figure 6.13. The share of different sources of energy during the evolution of technology. Year 2020 according to world energy conference.

In Figure 6.13 some data is given on the evolution of the different energy sources.

Now the problem of the present sources of free energy can be discussed in this context, beginning with man's first fuel, wood.

6.6 The Indirect Use of Solar Energy

6.6.1 The biosphere as Man's energy source for technology

The first technological source of energy for man was wood. This was his fuel for "space heating", for preparing food, and for light. The level of use has been estimated to be 0.3 kW per capita. Perhaps today the level is still about the same on the average and is important in the developing countries.

The burning of wood cannot be considered as a past chapter in the history of mankind, however, as there are more or less well founded proposals for the future utilisation of solar energy from plantations ("silviculture", that is, industrial wood growing).

One of the favoured scenarios of wood-use involves poplars, as one such tree needs $1\,m^2$ of space and after 12 years can be harvested and burned. During those 12 years the solar energy can be estimated as:

$$150\,W/m^2 \times (3.15 \times 10^7\,s/year) \times 12\,years = 57\,GJ.$$

It has been estimated that the efficiency of these solar plantations can reach 0.6%, that is, 3 times higher than the natural forest. This efficiency corresponds to an amount of harvested energy of approx. 340 MJ after 12 years from each square metre, $1.6\,kg/m^2 \cdot year$, although more recent studies give a higher value of $\sim 2.5\,kg/m^2 \cdot year$. After burning in a power station with an efficiency of 30%, $\sim 30\,kWh\,(el)$ (kWh(el) = kilowatt-hour electrical energy) of energy can be produced after 12 years. This corresponds to an energy flux of 0.3 W (optimistically, 0.45 W) of electrical energy per square metre.

For a region with 1,000,000 inhabitants with an electrical power consumption of 1 kW (el) per capita (the present level for industrial countries), a power station of 1 GW (el) is needed. The "solar plantations" must thus cover a surface of $3300\,km^2$. But in an industrialised country the population density is about 200 inhabitants per square kilometre or, for 1 million people, $5000\,km^2$. From this, $3300\,km^2$ is to be covered with poplar trees! The problems of ash, mineral fertilisers, protection against pests, and transport of 16,000 tons of wood per day have not even been mentioned.

The worldwide potential of the biomass, which can be used in power stations, has been estimated as 6 TW-years of electricity per year. Assuming an

6.6 The Indirect Use of Solar Energy

average harvest of 1 W/m², the surface of the "energy plantation" must be approx. 18×10^6 km², that is, more than the total land area devoted to the present worldwide agriculture. In this case too, not all the energy needs of Man can be covered by this kind of indirect solar energy supply, but only the 3 kW of thermal energy per capita.

Another proposal involves marine solar-energy plantations. The harvesting of kelp (large, brown seaweed) can produce a dry mass as fuel for a power station. The efficiency of converting the solar energy into dry biomass seems to be lower than that for the poplar forest. The productivity of 1 m² of sea is also much lower than on land.

It can be assumed that this kind of solar energy use will not be adapted in the future. It seems to be much more reasonable to use the harvested wood from the solar plantation on land, not for the production of electrical energy, but as a raw material for "industrial agriculture". About 1.5 kg per year per square metre of the harvested wood can be chemically hydrolised and transformed into sugars and other substances which can be used for the culture of microorganisms. These produce proteins which, taken along the appropriate production chain, can end up as human food. Under favorable conditions, 0.1 kg of proteins per 1.5 kg wood can be produced, which will supply a man's need for approx. two days.

6.6.2 Transformation of solar into kinetic energy: Wind

One of the oldest forms of man's technology is the transformation of wind into kinetic energy via the sails of ships or of windmills.

The total wind power on Earth is estimated to be some thousands of terrawatts, that is, approx. one-hundredth of the total solar energy flux. Some authors give a value of 10,000 TW for the total energy of wind on the Earth.

But there are difficulties involved in harnessing wind power; the greatest of these difficulties arise from the following:

- The wind is an extremely irregular phenomenon, therefore an energy storage system must be included.
- The velocity of wind changes dramatically, therefore a rather complicated system of regulation of the rotational speed of the generator is necessary.

The role of wind as a source of technological energy seems to be limited to a few regions of the world and is therefore only of local importance.

The prognosis for the USA resulting in 20 GW (el) by the year 2000 and 60 GW (el) for 2020 from this source seems to be rather optimistic (Figure 6.14).

Some efforts are being made in the field of power extraction from tides and oceanic waves, but the efficiency and real potential of these sources do not seem to be very encouraging. (Figure 6.14 gives some typical data.)

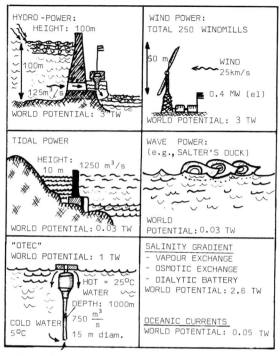

Figure 6.14. Transformed solar energy: hydro, wind, ocean thermal gradients (power station of 100 MW (el)).

6.6.3 Transformed solar energy: The kinetic energy of falling water

From the 120,500 TW of solar energy absorbed by the Earth, approx. 44,000 TW – that is, 36% – is used for the vaporisation of water. About 16.3×10^9 kg/s of water is involved in this process, the largest energetic process on the Earth:

$$\text{heat of vaporisation} \times \text{mass of water/sec} = \text{power}$$
$$2.7 \text{ MJ/kg} \quad \times \quad 16.3 \times 10^9 \text{ kg/s} \quad = 44,000 \text{ TW}.$$

Let it be assumed that all this water vapour has been transported to a height of 3000 m and there condensed. In a very simple way it can be construed that this amount of water falling from this height to the Earth's surface represents:

$$16.3 \times 10^9 \text{ kg/s} \times 3000 \text{ m} = 4.9 \times 10^{13} \text{ kg} \cdot \text{m/s}$$
$$= 480 \times 10^{12} \text{ J/s} = 480 \text{ TW}.$$

Limiting the calculation to the continents, the potential maximum power of falling water in the rivers is 2.8 TW. Another, probably better, value (1.2 TW) is more pessimistic and probably represents the true potential.

6.6 The Indirect Use of Solar Energy

The presently installed hydroelectric power stations represent a power of 0.15 TW (el), that is, 1/8 of the estimated potential of 1.2 TW. The power of a given plant is estimated as follows (see Figure 6.14):

Power = amount of water/s × height × efficiency.

For example, for a 100 MW (el) plant:

100 MW (el) $\cong$ 125 m³ water/s × 100 m height × 0.8 efficiency.

If one assumes that the collecting efficiency for rainfall is 50% and the annual rainfall is 0.8 m, then, for such a 100 MW (el) power station, a collection area of 9.8×10^9 m² is needed, that is, 9,800 km², or a square of ~ 100 kilometres on a side.

6.6.4 The "insignificant" form of solar energy: The heat of the oceans

The largest short-time storage place for solar energy is the tropical oceans. Here is the simple calculation:

- The upper layers of the tropical oceans have an average temperature of 25 °C (Figure 6.14).
- The lower layers in the same region are much colder – at 700 m depth the temperature falls to 5 °C.
- This temperature gradient of 20 °C can give a difference for heat extraction of, say, 10 °C. Then the heat capacity of one cubic metre of water equals:

$$1 \text{ m}^3 \times 10^3 \text{ kg/m}^3 \times 4.18 \text{ kJ/kg} \cdot \text{K} \times 10 \text{ K} = 41 \text{ MJ/m}^3.$$

The question arises: Can this convertible energy be transformed into the most useful form of energy for man, electricity? The answer is without doubt positive, but with reservations. The first point concerns the thermodynamic efficiency of such an energy transformation. The Carnot principle permits the calculation of the maximum theoretical efficiency (temperature in Kelvin):

$$\text{Efficiency} = \frac{\text{max. temp.} - \text{min. temp.}}{\text{max. temp.}} = \frac{298 - 278}{298} = 0.033.$$

This result shows that the efficiency of transformation of heat into electricity must be less than 3.3%. Power stations utilising this source are called OTEC's. There are some doubts as to whether such devices can do more than solve purely local energy problems.

There are some proposals for using the salinity gradient between oceanic and river water (e.g., Congo river with a potential of ~ 100 GW (el); see Figure 6.14).

6.6.5 The best forms of stored solar energy: Oil and coal

Wood is a short-term solar energy store. For the living tree the energy is stored (by transformation via photosynthesis) for about 10–15 years. As wood, the storage lasts about 1000 years. However, over hundreds of millions of years, solar energy has been stored as coal, gas, and oil in a high-energy form – high-energy because it is stored in a chemical form, the chemical bond being the well-ordered arrangement of the electrons in molecules.

The utilisation of this type of energy triggered the first industrial revolution in the eighteenth century, and there is no doubt that the exponential increase in the world's population in the twentieth century can be attributed to the large reserves of coal, and later oil, which began to be exploited at that time. This extends back into the Carboniferous Period, when these vast quantities of biomass were first laid down. In Table 6.6 it can be seen that the stored solar energy exists primarily in the form of coal, of which the total amount of world reserves equals 7.4×10^{15} kg, that is, $\sim 2.20 \times 10^{23}$ J. This amount of coal contains the same amount of energy as the Earth receives from the Sun in ~ 23 days. However, taking into account only that part of solar energy transferred by photosynthesis, 92 TW, the total coal reserves then represent approx. 75 years' equivalent of the total biosphere's net production. The oil and gas reserves are five times smaller than those of coal.

The present consumption of fossil fuel – that is, coal, oil and gas – equals 9.5 gigatons/year, which is equal to 3.0×10^{20} J per year. The mean energy flux is 9.5 TW. There are also other sources of energy (e.g., wood).

The present world population of 4.8 billion consumes approx. 2.2 kW of total energy per capita. This energy flux of 11 TW corresponds to:

$\sim 7\%$ of the total solar flux involved in photosynthesis;

0.008% of the total solar energy being absorbed by the Earth.

There is no doubt that by utilising sources of energy for technology which total such a large proportion of the total energy flux, man controls a parameter which has a significant influence on the world energy balance.

In the past, the relative shares of various kinds of energy have reached a peak and then fallen off in favour of other energy sources. The reasons for the changeover were not always economical or a question of availability. Sociopolitical and technological requirements also played a role. This was particularly true when large transport paths would have been necessary, up to, say, half the Earth's circumference, to enable supply to satisfy demand.

More recently, environmental aspects have become more important in the choice of the best fuel for a particular purpose. There is concern over ash production and air pollution by CO_2 and sulphur oxides linked to the combustion of fossil fuels.

A fairly characteristic phenomenon is the resistance to despoliation of the Earth's surface, particularly in those countries where coal is produced by strip-mining. Taking the example of the United States in 1975, the strip-mining of coal typically gives 420 kg coal per square metre. This amount of coal contains about 12 GJ of energy and, assuming coal must be burned for 12

6.6 The Indirect Use of Solar Energy

Table 6.6. Stored Energy on the Earth

Fuel	Remark	Global resources in 10^{15} kg	Terawatt-years	Total solar terrestrial input (1 solar day = 300 TW-year). in solar days
Fossil				
Oil (1 kg = 42 MJ)	World	1.35	850	
Gas (1 kg = 42 MJ)	World	0.87	550	4.7
Methane (1 m³ = 40 MJ)	Geopressurized (USA only)		1500	~5
Coal (1 kg = 28 MJ)	Total reserve resources	7.4 1.7	7000 1600	23 ~5
Peat (1 kg = 20 MJ)	(USA only)	0.4	540	1.8
Nuclear fission				In solar years (1 solar year = 109,000 TW)
Uranium (1 kg = 86 TJ)	In Earth's crust: ~3 ppM Extraction of 1/1000 only	0.7	1.9×10^9	15,000
Thorium (1 kg = 86 TJ)	In Earth's crust: ~12 ppM Extraction of 1/1000 only	2.8	7.9×10^9	65,000 ―――― 80,000
Nuclear fusion				
Deuterium (1 kg = 275 TJ)	In Oceans: 16 ppM Extraction of 1/100 only	2.3 0.2	1.9×10^9	15,000
Lithium (as source for Tritium)	In crust: 30 ppM	~0.2	1.9×10^9	15,000 ―――― 30,000

ᵃ 1 Terawatt-year = 1 TWy = 14 million barrels oil per day during year
 = 1.06×10^9 tons coal equivalent
 = 0.79×10^9 tons crude oil
 = 3.15×10^{19} J

days to yield an energy flux of 12 kW per capita (average consumption in USA), it can be calculated that during the average life of an American (say 70 years) a surface of 2200 m² of strip-mined coal is needed. It must be remembered that for a population density of ~200 people/km² only 5000 m² per capita is available to provide a total life's needs.

In the assumed conditions, the energy flux per square metre of coal mining is 12 kW, that is, a thousand times more than the optimistic estimation for direct solar energy (see Section 6.7) and approx. 10,000 times more than the average energy flux in photosynthesis (see Section 7.3.2). It can also be seen that this 420 kg of coal, containing 12 GJ of stored solar energy, was accumulated over a time span of some thousand years of photosynthetic action.

But coal mining includes one more paradox, to explain which consideration must be given to some aspects of nuclear energetics.

In the USA, and probably also in other countries, coal contains some small amounts of the heaviest naturally occurring elements, thorium and uranium. As can be seen in Table 6.6, both elements are present in the Earth's crust in the following concentrations: thorium 12 ppM, uranium 3 ppM. There is nothing surprising in the fact that coal, too, being chemically active and in contact with rocks and water, has absorbed some of these elements. The average concentration in coal is thorium (2 ppM) and uranium (1 ppM). The ashes of some European coal include 30–50 ppM of both elements.

The amount of energy contained in these small amounts of trace elements, when "burned" in nuclear fission, in the 420 kg of coal coming from the 1 m² of strip-mine, can be calculated. It must only be known that the "burning" of uranium and thorium in a breeder reactor releases the following energy:

$$1 \text{ kg uranium or thorium} \sim 8.6 \times 10^{13} \text{ J} = 86 \text{ TJ}.$$

In 420 kg coal the 3 ppM of fissionable elements represents the following amount of energy:

$$420 \text{ kg coal} \times (3 \times 10^{-6}) \times (86 \times 10^{12} \text{ J/kg}) = 10^{11} \text{ J} = 100 \text{ GJ}.$$

Summarising: ~1 m² of strip-mine yields 420 kg of coal, which can be chemically burned to give 12 GJ of energy. The same 420 kg of coal ends up as ashes, which contain 1.3 grams of thorium and uranium, which in a breeder reactor could be used to give 100 GJ of energy, that is, 8 times more than the chemical energy contained in the coal!

There is one more significant problem connected with the burning of coal: The average sulphur content is 3 %. To obtain 1 gigajoule of energy, 35 kg of coal, which contain 1 kg of sulphur, must be burned. In some areas of the United States with severe environmental restrictions, the limits of sulphur that may legally be released have been set at 0.1 kg of sulphur per 1 GJ of energy produced. Processes of desulphurisation exist which reduce the amount released to the air, but, as a result, solid waste is produced containing sulphur in amounts equivalent to one-third of the coal burned. Burning 1×10^9 tons of coal per year, the sulphur-containing wastes are 10^8 tons – a tremendous amount. The processes of the desulphurisation are expensive.

The same amount, 1 GJ, of energy can be produced from 35 kg of coal (which also produces 3 kg of ash and 3 kg of sulphur-containing waste) or from 12 mg of thorium and/or uranium. The radioactive fission products formed in the production of this 1 GJ amount to 12 mg of waste and, after solidification, 1 g of solid radioactive waste.

Of course, the 1 g of highly radioactive waste is much more dangerous than the ashes and solidified sulphur; nevertheless, this comparison reveals the problem of waste production by the different processes (see Section 6.10.4).

6.7 The Direct Technological Use of Solar Energy

6.7.1 The simplest way: Space heating

Man needs a lot of ambient heat to maintain subtropical conditions in his direct environment. In the highly industrialised countries, the energy flux for this purpose is 1–2 kW per capita in the coldest period. The space requirement for one man can be assumed to be equivalent to 30 m² of surface for dwelling, offices, hospitals, schools, shops, and industry. This corresponds to an energy flux of 100 W/m² in the coldest period.

The average solar flux on the Earth's surface is also 120 W/m², about that needed for space heating.

There are several difficulties in this kind of direct use of solar energy:

– The maximum space heating requirement comes at the time of the minimum solar energy flux.
– Space heating is also needed at night.
– A rather large proportion of the population lives in multistory dwellings, and the corresponding free surface for solar absorption is smaller.
– Solar energy collectors have a rather low efficiency and cannot absorb more than 1/3 of the total received flux; also, the storage of absorbed energy is rather expensive.
– Solar collectors are not simple to construct. Their fabrication depends on technology, which itself requires energy. It has yet to be shown that this energy requirement is not too high in relation to the solar energy to be utilised.

Clearly, in some regions of the world direct solar heating of space is the best solution, but this cannot be the general solution to mankind's energy problems. At least it can be claimed that it is the cleanest, simplest, and most direct way of using solar energy.

6.7.2 Solar energy converted into electricity on the Earth's surface

The average solar flux on the Earth's surface equals 49% of the total solar flux (Figure 6.3) or 168 W/m^2. At noon this can rise to 600–800 W/m^2.

The transformation of solar flux into electricity is possible in several ways (Figure 6.15):

- by means of semiconductors in which the photons move electrons; that is, the photons are transformed directly into electrical energy (a very good candidate: hydrogenated amorphous silicon);
- by means of chemical photo-reactions in which the photons break the chemical bonds; that is, the photons are transformed into chemical energy;
- by means of transformation to heat and then into the movement of electrons (electrical energy) in so-called thermionic devices;
- by means of transformation to heat at a relatively high temperature and then through a steam turbine into mechanical energy and finally into electrical energy (see Table 6.7).

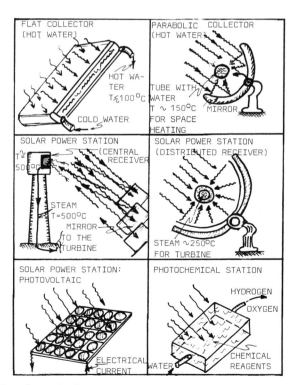

Figure 6.15. Transformed solar energy: hot water, thermal power station, photovoltaic power, photochemical station.

6.7 The Direct Technological Use of Solar Energy

Table 6.7. Transformation of Solar Energy into Electrical Energy

Primary energy	Primary/ secondary transformation	Electricity production	Total efficiency	Surface[a] m^2/kW (el)	Capital costs US$/kW (el)
Photons, visible light	*Photovoltaic* ("solar cells") *Semiconductors* – Silicon – Gallium arsenide – Silicon-hydrogenated, amorph.	Direct, but low voltage No storage	<0.28 (theoretical ~0.40)	22 m^2	15,000 $20,000/kW (el), reduction to $1200/kW is possible
	Photochemical Storage of chemical agents, if needed transformation into electrical energy by means of batteries or by burning.	Indirect Inherent storage	≪0.30	>11 m^2	$600/kW of chemical energy (?) (very optimistic!)
Thermal energy, heat	*Thermoionic* Gallium-selenide	Direct, low voltage no storage	≪0.20	>17 m^2	Very expensive ?
	Solar power tower with heliostats "CRS"; central receiver system. $T_{max} = 530\,°C$ (Fig. 6.14)	Mirrors, (heliostats) and power tower: steam with temp. ~550 °C steam turbine generator.	At midday 0.18 Annual average $50\,\frac{W(th)}{m^2}$	60 m^2 [b]	$70–145/$m^2$ $1000–2000/kW (el) Remark: In USA, 1981, solar power test station ~5 MW(th) In Albuquerque: 10 MW (el) in Barstow. Also in Almeria (Spain): 1 MW (el)
	"DCS"; distributed collector system. $T_{max} = 250\,°C$ (Fig. 6.14)	Heat storage, in molten salt, then production of steam with temp. 250 °C steam turbine generator	<0.25	13 m^2 [b]	Payback of energy (in northeast USA) ~15 years.

[a] All data for daylight with 300 W/m^2.
[b] A solar power station with 1 GW (electrical) needs a surface from 9 to 22 km^2 for the solar collectors.

6.7.3 The extraterrestrial conversion of solar into electrical energy

The technologies for converting solar into electrical energy on the Earth's surface described above suffer from the following restrictions:

- The mean flux of solar energy over a long period is not more than $\sim 170 \text{ W/m}^2$.
- The efficiency of conversion is low, 0.15, or 25 W/m². The rest of the energy is converted to heat and, in the best case, cooling must additionally extract 145 W/m².
- The energy flux varies from zero during the night to 800 W/m² at midday and in good weather, which causes problems of energy storage and transport.
- The cost of a unit of the Earth's surface cannot be neglected, particularly in the future, when the population density of the world will increase.

A rather reasonable solution to these difficulties can be achieved by putting the solar cells far outside the terrestrial atmosphere, for example, at a stationary orbit 36,000 km high. In this case, the solar energy photovoltaic transformation can be characterised as follows:

- The mean flux of solar energy = 1367 W/m² (over the total solar spectrum), in contrast to ~ 168 W/m² on the Earth's surface.
- The efficiency of conversion is low and the cooling is difficult because of the lack of an atmosphere.
- The average energy flux is fairly constant and becomes zero only when passing into the Earth's shadow, for ~ 1 hour/day.
- The cost of collection at the Earth's surface can be reduced, since the antenna for receiving the energy from such a station can accept a flux density one order of magnitude higher than a surface solar cell.

Of course, another technological problem has to be solved: the transport of the electrical energy from the orbital solar station 36,000 km above the Earth's surface. This appears to be soluble by means of a microwave system (Figure 6.16).

Recently there has been a lot of criticism of the concept of an orbiting solar power station. The detailed calculations of the amount of energy required to construct such a power station and the terrestrial receivers give a rather doubtful result. This means that the amount of energy put into construction and maintenance is larger than that obtained from the system in a reasonable period of time.

Probably such studies exaggerate the problem and miss some important objectives. A fuller investigation of the problems and possibilities is very important, but the first negative results cannot be ignored.

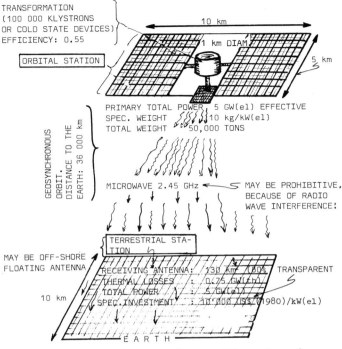

Figure 6.16. Solar power from satellites: a dream?

6.8 Technological Use of Non-solar (Nuclear) Energy

6.8.1 The heaviest elements: The gift of the supernova

Approximately five billion years ago, in this region of the Milky Way, a supernova explosion influenced the evolution of the stellar matter.

Supernovae seem to be the only possible source of the heaviest elements, thorium and uranium (Chapter 3). Some of these supernova products have a half-life on the order of the age of the Solar System:

	Half-life (gigayears)	Average concentration in the Earth's crust (part per Mega (ppM))
Solar System	Age ~5	
Uranium-238	4.51	4.0
Uranium-235	0.71	0.29
Thorium-232	13.9	12.0
Potassium-40	1.27	4.0

These nuclides are, of course, unstable. The heavy nuclides are alpha-unstable, while the daughter products and potassium-40 are beta-unstable.

By a very complex geochemical evolution most of these radioactive nuclides have been accumulated in the thin layer (25 km thick) of the Earth's crust (Chapter 4).

6.8.2 Geothermal energy results from the nuclear decay of radionuclides

The decay of the long-lived nuclides uranium-238, uranium-235, thorium-232, and potassium-40 results in the production of heat in the following amounts (watts per gram of nuclides):

Uranium-238	9.5×10^{-8} W/g
Uranium-235	0.44×10^{-8} W/g
Thorium-232	2.6×10^{-8} W/g
Potassium-40	2.9×10^{-8} W/g

The total energy flux resulting from these decays is equal to 1×10^{-9} W/kg of the Earth's crust ($\sim 2 \times 10^{-9}$ W/kg of granite).

If it is postulated that the crust is 25 km thick, then from each square metre of Earth's surface an energy flux of ~ 0.06 W/m² is produced:

$$(1 \times 10^{-9} \text{ W/kg}) \times (2700 \text{ kg/m}^3) \times (25{,}000 \text{ m}) = 0.06 \text{ W/m}^2.$$

The deeper layers of the Earth produce roughly the same energy flux. For the total Earth this equals 57 TW. The total energy coming from the spontaneous decay of radionuclides is only 1/2200 of the total solar energy absorbed by the Earth. Additionally, a flux of 9 TW comes from primordial heat.

In one or two almost unique regions, this geothermal flux is so large that it might be feasible to harness this heat. Knowing the physical properties of the average rocks, density 2700 kg/m³, specific heat 0.8 kJ/kg·K, and thermal conductivity ~ 2 W/m·K, it can be estimated that the average thermal gradient due to this flux equals 33 m for every degree Kelvin (in volcanic rocks only 2 m/K, in frozen rocks ~ 500 m/K).

There are two possible ways of harnessing the geothermal energy:

- exploitation of the constant flux, that is, ~ 1 W/m² in the "hot" regions, and
- extracting the sensible heat in the rocks and cooling them by some hundred degrees.

For the first case, the extraction and use of the geothermal heat for a power station with 50 MW (el) for approx. 30,000–50,000 inhabitants would require a surface area of 350 km².

Under one square kilometre of surface to a depth of 10 km – that is, in 10 cubic kilometres of the Earth's crust, assuming a temperature rise of 100 K –

6.8 Technological Use of Non-solar (Nuclear) Energy

2×10^{18} J of energy are stored. This corresponds to 68 million tons of coal. The exploitation of this geothermal heat, even with an efficiency of 10%, over 50 years, gives enough for a power plant of 125 MW (el). This is enough for approx. 20,000 people in an industrialised nation. It is believed by some that geothermal energy is not easy to exploit, but if it can be mastered, then here is a clean energy source. Unfortunately this is not so. Natural hot waters, it must be remembered, contain 0.5–1% non-dissolved gases, for example, carbon dioxide, sulphur, hydrogen, and, in lower amounts, methane, ammonia, and traces of arsenic. In the water are also significant amounts of silica and carbonates. The natural steam reaches the surface with a high noise level and the low efficiency of the transformation of heat into electricity causes a thermal pollution of the environment.

In Iceland, the paradise of geothermal energy, the total thermal power foreseen is 1 GW (thermal) after drilling down to 2000 m. For the future, a 60 MW (electrical) power station is planned.

Even in volcanic Japan, the most optimistic prognosis for the use of geothermal power is about 5 GW (el) of power stations in the year 2000, corresponding to about 1% of the total installed power.

The contribution of geothermal energy is, however, not limited to its use as a new source of technological power. The most important role played by radioactive decay of uranium and thorium is in volcanic activity, seismic activity, and in the movement of continents. Here the influence is felt in the overall evolution of the terrestrial climate, the biosphere, and of mankind (see Chapter 6.2.7).

6.8.3 The fission of the heavy nuclides is one of the most abundant terrestrial energy sources

Geothermal energy based on spontaneous radioactive decay consumes, over billions of years, only a very small amount of the available nuclear sources, and at very low specific powers.

Another process exists which occurs very quickly, with a high specific power – extracting the nuclear energy from the nuclei of uranium and thorium atoms, that is, controlled nuclear fission. In Table 6.8 the processes are shown side by side (see also Figures 6.17 and 6.18).

As we have seen, the heavy elements (Chapter 3) are produced only by a nuclear process in which energy is consumed. All elements heavier than iron are the product of such energy-consuming processes. These nuclides are unstable, that is, they can decay and release free energy. But the probability of such spontaneous decay is very low.

Some of the decay processes can be induced by means of particles or quanta. A special case is when these same particles are also emitted in the process of the

Table 6.8. Heavy Nuclides Decay

Character of the process	Alpha-decay, spontaneous	Fission process	
		spontaneous	induced by neutrons
Mean energy released per nuclide	5 MeV	200 MeV	200 MeV
Half-life for the process	10^9 years	10^{15} years	10 years (in a reactor)
Decay products	1 α-particle and 1 heavy nuclide	2 or 3 neutrons and 2 mean mass nuclides	
Specific power of energy released in: – relative numbers – W/kg	$\frac{5}{10^9} = 5 \times 10^{-9}$ 10^{-4}	$\frac{200}{10^{15}} = 2 \times 10^{-13}$ 10^{-8}	$\frac{200}{10} = 20$ 10^6

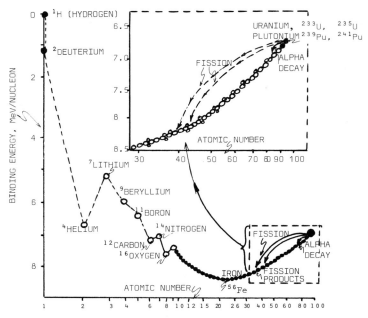

Figure 6.17. The release of free energy from alpha decay and fission of heavy nuclides.

6.8 Technological Use of Non-solar (Nuclear) Energy

induced decay, which can be termed self-catalysing; for example, a non-self-catalysing process:

$$\text{heavy unstable nuclide} \xrightarrow[\text{(very long time)}]{\text{spontaneous decay}} \left\{\begin{array}{l}\text{decay} \\ \text{products}\end{array}\right\} + \text{energy};$$

and a self-catalysing or chain process (energy omitted):

$$\text{particle} + \text{heavy nuclide} \xrightarrow[\text{(very short time)}]{\text{induced decay}} \begin{array}{l}\text{decay} \\ \text{products}\end{array} + \text{particles}$$

A more interesting case is when the number of particles produced by induced decay is greater than 1, or even greater than 2. This is an "avalanche" process. Such a case can be observed by the induced decay, or rather induced fission, of uranium-235:

$$\text{neutron} + \text{U-235} \xrightarrow[\text{(very short time)}]{\text{induced fission}} \left\{\begin{array}{l}\text{fission} \\ \text{products}\end{array}\right\} + 3 \text{ neutrons}$$

2 free neutrons

In macroscopic amounts, the fission of atomic nuclei releases the following amount of energy (in the case of a total breeding process, that is, the transmutation of uranium-238 into plutonium-239 and thorium-232 into

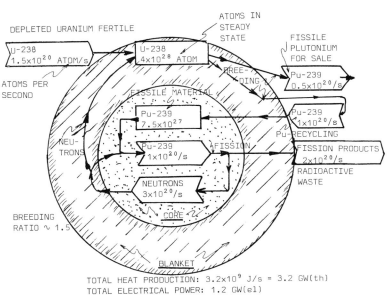

Figure 6.18. Diagram of a fission breeder reactor working on a uranium-plutonium cycle. Initial requirement of structural materials: 6×10^6 kg radioactivity after shutdown (bequerel); after 10 sec: 5.5×10^{20} Bq; after 1 year: 1×10^{19} Bq; after 10^6 years: 1×10^{13} Bq.

Table 6.9. Processes in an Advanced Fast Breeder Power Reactor with Electrical Power ~1.2 GW (el) (see Figure 6.18)

Parameter	Unit	Value
Amount of Plutonium in the core of reactor	kg	3000
Number of Plutonium atoms $(6 \times 10^{23}\,\text{atom}/0.239\,\text{kg})$	atom	7.5×10^{27}
Number of atoms fissioned during 1 second	atom/s	1×10^{20}
Results of fission of 1 atom:		
Energy, total	MeV	200
Products; fission fragments, neutrons	atom	2
	neutron	3
Energy released in the core during 1 second 1 Joule/$(6 \times 10^{18}\,\text{eV}) \times (1 \times 10^{20}\,\text{fission/s})$ $\times (2 \times 10^{8}\,\text{eV/fission})$	J/s	3.2×10^{9}
Power (thermal) of this reactor	W (thermal)	3.2×10^{9}
Mass of Plutonium fissioned per second $(10^{20}\,\text{atoms/s}) \times (0.239\,\text{kg}/6 \times 10^{23}\,\text{atom/mol})$	kg/s	4×10^{-5}
Mass of Plutonium fissioned per day	kg/day	3.45
Number of neutrons released during fission	neutron/s	3×10^{20}
Number of neutrons used for fission	neutron/s	1×10^{20}
Number of neutrons captured by parasitic nuclei	neutron/s	0.5×10^{20}
Number of neutrons captured by Uranium-238 and transformed to Plutonium-239	neutron/s	1.5×10^{20}
Amount of Uranium-238 transformed to Plutonium-239 during 1 second	kg/s	6×10^{-5}
Amount of Uranium-238 transformed to Plutonium-239 during 1 day	kg/day	5.1
Net increase of the Plutonium in the breeder during 1 day (5.1–3.2)	kg/day	1.9
Amount of days needed to produce Plutonium for a new core with 3000 kg Plutonium (3000 kg)/(1.9 kg/day)	day	1580
	years	~4.3

uranium-233):

1 kg of uranium (or thorium) fissioned releases 86×10^{12} J = 86 TJ of free energy. (Note: this figure makes no allowance for material or energy losses.)

Table 6.9 gives the most important data for a fast breeding reactor. In Figure 6.18 the outline of such a reactor is shown.

Simplified calculations of the world resources of fissionable nuclides are summarised in Table 6.10.

6.8 Technological Use of Non-solar (Nuclear) Energy

Table 6.10. Resources for the Carriers of the Fission Energy: Uranium, Thorium

Parameter	Unit	Value
Surface of Earth	m^2	510×10^{12}
Surface of continents	m^2	150×10^{12}
Thickness of continental crust	m	20,000
Volume of continental crust	m^3	3×10^{18}
Specific weight of rocks	kg/m^3	2500
Mass of continental crust	kg	7.5×10^{21}
Uranium content in rocks	ppM by weight	3
Thorium continent in rocks	ppM by weight	12
Mass of Thorium and Uranium in continental crust	kg	1.1×10^{17}
Ratio of useful recovery (arbitrary)	%	3
Mass of Thorium and Uranium recoverable	kg	3.4×10^{15}
Specific energy content in Thorium and Uranium, after breeding	$\dfrac{J}{kg}$	8.6×10^{13}
Energy content of this Thorium and Uranium, with efficiency ~ 0.6	J	1.7×10^{29}
World population in the future	people	8×10^9
Primary energy flux in the future per capita (see Chapter 8)	kW/capita	8
Primary energy flux in the future for world population per year	J/year	2×10^{21}
Uranium + Thorium burned per year	kg/year	3.9×10^7
The resources of Thorium and Uranium will be exhausted after $(1.7 \times 10^{29} \text{ J})/(2 \times 10^{21} \text{ J/year})$	year	84,000,000

Other Resources of Uranium

Geological reservoir	Concentration ppM Uranium	Amount kg Uranium
Vein deposits	3000	2×10^9
Fossil placers, sandstones	1000	6×10^{10}
Volcanic deposits	100	1×10^{12}
Shores (black, phosphates)	10	5×10^{14}
Granites	3.5	2×10^{15}
Ocean water	~ 0.0033	5×10^{12}

6.8.4 Fusion: The second coming of nuclear energy

Most energy in the Universe seems to arise in the very slow "burning" of hydrogen (see Figure 6.19):

$$4 \text{ hydrogen nuclei} \xrightarrow[\text{very slow}]{\text{fusion}} 1 \text{ helium nucleus} + \text{energy}.$$

The full process is:

$$4 \text{ kg of hydrogen} \rightarrow 4 \text{ kg/helium} + 2.7 \times 10^{15} \text{ Joules} \cong 2.7 \text{ PJ}.$$

The same, on a macroscopic scale:

$$4 \text{ kg of hydrogen} \rightarrow 4 \text{ kg/He} + 2.7 \times 10^{15} \text{ Joules} \cong 2.7 \text{ PJ } (P = \text{Peta} = 10^{15}).$$

For the sake of illustration, the relationship between fission and fusion can be shown by the ratio of the specific energy contents:

$$\frac{\text{energy released in fission of } U}{\text{energy released in fusion of } H} = \frac{0.086 \text{ PJ/kg}}{2.7 \text{ PJ/kg}} = \frac{1}{31}.$$

This "simple" process of the fusion of four hydrogen nuclei is not very probable. The real process goes by intermediate steps in which no more than

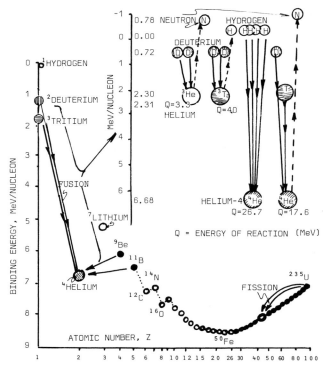

Figure 6.19. Fission and fusion versus the nuclear binding energy per nucleon.

6.8 Technological Use of Non-solar (Nuclear) Energy

two particles can collide at once. For example:

$$\text{proton} + \text{proton} \rightarrow (\text{di-proton}) \rightarrow \begin{cases} \text{energy} \\ \text{deuterium} \\ \text{neutrino} \\ \text{positron} \end{cases}$$

When it is taken into account that the deuterium is a compound object containing one proton and one neutron, it can be seen that this fusion process involves a transmutation of one proton into one neutron (p^+: proton; n^0: neutron; e^+: positron; v^0: neutrino):

$$p^+ \longrightarrow d^+ \text{ (deuterium)}.$$

$$p^+ \xrightarrow[\text{beta-decay}]{\beta} e^+ + v^0 + n^0$$

Table 6.11. Thermonuclear Devices: Natural and Technological

Fusion "machine"	Thermonuclear devices		
	Sun	Fusion reactor	
Primary process	Proton + Proton → → Deuterium	Remark: Tritium from breeding reaction: $^6_3\text{Li} + ^1_0\text{n} \rightarrow ^3_1\text{T} + ^4_2\text{He}$	
Secondary process	Deuteron + Deuteron → → Helium-4	Deuterium + Tritium → → Helium-4 + Neutron	
Energy per reaction	$Q = 26.7$ MeV	$Q = 17.6$ MeV	
Elementary interaction	For the Proton reaction: the *weak* interaction	For the Deuterium-Tritium reaction: the *strong* interaction	
Time constant for this interaction (relative)	10^{-6}	1	
Specific power $\dfrac{\text{J/s}}{\text{kg}}$	2×10^{-4}	$\sim 10^8$	
Specific power (relative)	1	$\sim 10^{12}$	
Confinement mechanism	Gravitational confinement because solar mass $\sim 2 \times 10^{30}$ kg	Magnetic confinement	Inertial confinement
		Strong magnetic field	Laser induced micro-explosion / Implosion induced by macro-explosion
Device (reactor or bomb)	Stars only!	Controlled thermonuclear reactor	Thermonuclear bomb

This process is the β-plus process (see Chapter 1). The beta processes are controlled by the "weak interaction", and these "weak" processes are very slow. This is the reason for the extended burning of the Sun over a long time period (Chapter 3).

The technological sources of energy must be of high specific power, about 1 kW/kg (for illustration, the human body produces 1 W/kg). Table 6.11 shows the relationship between the natural fusion device, the Sun, and a hypothetical fusion reactor.

It is worthwhile repeating that the fusion reactor, being the target of recent development work in numerous laboratories around the world, must have a specific power of approx. 10^{12} J/s · kg, that is, a thousand billion times greater than that of the Sun. In Table 6.12 some very rough data on the resources of the carriers of the controlled thermonuclear reactors are summarised. Figure 6.20 gives a flowchart of a controlled thermonuclear power reactor of approx. 1 GW (el).

Table 6.12. Resources of the Carriers of Fusion Energy: Deuterium

Parameter	Unit	Value
Surface of Earth	m²	510×10^{12}
Surface of oceans	m²	360×10^{12}
Ocean depth, average	m	3800
Volume of oceans	m³	1.4×10^{18}
Specific weight of water	kg/m³	1000
Mass of oceanic water	kg	1.4×10^{21}
Content of hydrogen in water	kg/kg water	0.111
Mass of hydrogen in world ocean	kg	1.55×10^{20}
Deuterium concentration in hydrogen	ppM by weight	149
Mass of Deuterium in world ocean	kg	2.3×10^{16}
Useful recovery of Deuterium (arbitrary)	%	10
Mass of recoverable Deuterium	kg	2.3×10^{15}
Specific energy content of Deuterium (by fusion, roughly)	J/kg	5×10^{14}
Energy content of recoverable Deuterium	J	1.1×10^{30}
Primary energy flux in the future for the world population of 8 billions with 8 kW/capita	J/year	2×10^{21}
The resources of recoverable Deuterium will be exhausted after $(1.1 \times 10^{30})/(2 \times 10^{21})$	year	570,000,000

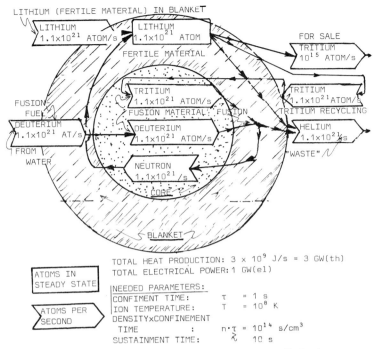

Figure 6.20. Diagram of the controlled thermonuclear (fusion) reactor. Initial requirement of structural material: 25×10^6 kg radioactivity after shutdown (bequerel); after 10 sec: 1.1×10^{20} Bq; after 1 year: 1×10^{19} Bq; after 1 million years: 7×10^{12} Bq.

6.9 Are There Other Sources of Energy?

The problem can be formulated as follows. The release of free energy is possible in a number of ways thanks to the four elementary forces: strong (nuclear), weak, electromagnetic, and gravitational. The specific energy release, that is, the energy released per nucleon, varies very significantly, from less then 1 eV up to half a billion eV (see Table 6.13).

At present the technology of energy extraction is based on only a portion of the theoretically possible methods of energy release. Even the use of the nuclear fusion of light nuclides is today only possible – that is, for a short intensive period – in a bomb. The continuous or quasi-continuous extraction of energy from this source using a thermonuclear reactor may become feasible only in the next century.

The best ratio for the extraction of free energy from a minimum amount of matter is given in some reactions connected with two important processes:

– annihilation of koino- and anti-matter (see Chapter 1), and
– gravitational collapse of matter (see Chapter 1).

Table 6.13. Primary Sources of Free Energy

Energy		Elementary interaction				Machine
In unit of mass	eV per nucleon	Strong	Weak	Electromagnetic	Gravitational	
					Rotational impulse	Earth?
10^{-9}	1			Atmospheric heat difference	Hydropower ($\sim 10^{12}$)	Hydropower station *** Solar power tower ***
10^{-8}	10			Fossil fuel (25×10^6)		Thermal machines (boiler, diesel)***
10^{-7}	10^2					
10^{-6}	10^3					
10^{-5}	10^4					
10^{-4}	10^5		Radioactive decay			Radioactive battery ***
10^{-3}	10^6	Fission (11.5) Fusion (0.375)				Fission reactor *** Fusion reactor (Contr. thermonuclear reactor) **
10^{-2}	10^7					
10^{-1}	10^8					
		Annihilation of antimatter (50% efficiency) (0.022)			Gravitational collapse (0.011)	Annihilation reactor?
1 ··	10^9					Black hole reactor?

*** Exists.
** Probably will be constructed in the next century.
? Possible, but probably not realistic.

In parentheses is the amount of matter (in kg) needed for extraction of free energy of 10^{15} J (1 PJ).

At the risk of repetition, the following points can be made:

– The annihilation of matter on the Earth, which is made of koino-matter, has been observed only in laboratories. The reason is that no free antimatter has ever been observed in nature. The spontaneous emission of antielectrons (positrons) in the beta-plus decay or emission of anti-neutrinos in beta-minus decay is neglected here. The anti-particles observed in laboratories have been synthesised with the aid of very large accelerators with rather large energy requirements. The laboratory-scale annihilation process has, therefore, a very negative energy balance; it is a sink, not a source, of energy. Only the discovery of significant amounts of anti-matter coming from outside, from cosmic space, can make the extraction of energy from this source possible. The chances of such an event are very small, if not zero.

- The gravitational collapse is a process which has been observed neither in the laboratory nor even in cosmic space. The process can only be deduced from the theory of general relativity. Some cosmic objects are suspected to be gravitationally collapsed stars – black holes – but there is no consensus about the true nature of these objects. The existence of very small collapsed objects of 10^{15} kg has also been postulated. Some speculations have even been made concerning the extraction of energy from such collapsed objects. All these are very far from reality and must therefore, at present, be classified under science fiction.

There are, however, still other possibilities for a "feasible" energy source. The following examples illustrate this:

- a thermodynamic machine using a temperature difference between the relatively hot terrestrial surface, with a temperature of ~ 300 K, and cosmic space, with a temperature of 3 K;
- the rotational momentum of the Earth equals 2.1×10^{29} J. The extraction of 60 TW (6×10^{12} W) of power over one million years corresponds to a total extracted energy of 1.9×10^{27} J. This is only 1 % of the total rotational momentum.

Both ideas seem attractive at first glance, but after a short consideration it must be clear that both are beyond the bounds of foreseeable possibility.

6.10 Energy Production as a Source of Dangerous Waste and Environmental Problems

6.10.1 Energy production and nonradioactive waste materials

The real difficulties with the present civilisation lie not in limited resources but with waste and pollution problems.

The types of waste produced by energy devices are numerous. The following kinds may be considered:

- material waste, e.g., carbon dioxide and sulphur dioxide from fossil fuel, or radioactive waste from nuclear power;
- thermal waste, that is, the low-grade energy flowing to the sink, e.g., in the cooling towers of power plants;

Table 6.14. Wastes from Energy Data. All Data Calculated for 1 GW Electrical Energy (efficiency $\sim 33\%$) over 30 years $= 2.9 \times 10^{18}$ J (ther) $= 2.9$ EJ (ther)

Primary source of energy on the Earth	Carrier of energy	Heat waste of the energy transformation processes	Material waste from production	Specific properties of the material waste
Solar	Massless photons on a surface of 60 km^2	3 times more than the converted energy, but does not influence the heat balance of the Earth	Only structural material, which can be fully recirculated	No natural waste; only "waste" is the surface of the Earth covered by the solar energy convertors
Fossil fuel	Coal, Oil Coal: 8.3×10^{10} kg Oil: 5.8×10^{10} kg	1.5 GW thermal ($\eta = 0.40$)	For coal: 3×10^{11} kg of carbon dioxide, 1.2×10^{10} kg of ashes, 3×10^{10} Bq in air, 5.8×10^9 kg sulphur dioxide	Unknown impact of carbon dioxide, sulphur dioxide, nitrogen oxide, ashes with radioactive substances
Fission	Uranium, Thorium: 4×10^4 kg	1.5–2 GW thermal	34,000 kg of radioactive waste $+ 10^6$ kg of irradiated structural material	Highly radioactive, very long-lived, very dangerous material 1×10^{17} Bq after 1000 years
Fusion	Deuterium + Lithium: 1×10^4 kg	1.5–2 GW thermal	1100 kg of Helium + 10^6 kg of irradiated structural material	Highly radioactive, long-lived, dangerous, material 1×10^{15} Bq after 1000 years

— the surface waste, that is, the despoliation of the Earth's surface. A short review is given in Table 6.14.

The material waste of the fossil fuel sources is clearly very high. Over a period of 30 years a coal-fired power station of 1 GW (el) produces 3×10^{11} kg of carbon dioxide, 5.8×10^9 kg of sulphur dioxide, and 1.2×10^{10} kg of ash.

This power station provides for the energy needs of a population of a quarter of a million people, but, over the past 30 years, the waste production has been terribly high. The amount of waste produced per capita during the lifetime of the average man (i.e., over 75 years) is:

carbon dioxide	3×10^6 kg =	3000 tons
sulphur dioxide	6×10^4 kg =	60 tons
ash	1.2×10^5 kg =	120 tons

6.10 Energy Production as a Source of Dangerous Waste and Environmental Problems

On taking the amounts relative to the mass of the human body (70 kg):

    carbon dioxide   42,000 body masses,
    sulphur dioxide   800 body masses,
    ash   1700 body masses.

From this it can be seen that man's impact on the environment in this respect is significantly large.

A nuclear fission power station appears to have a much smaller influence on the environment. The material wastes per capita (expressed in body mass terms) over 75 years are:

- radioactive waste $\sim \frac{1}{220}$ of the mass of the human body;
- radioactive structural material $\sim \frac{1}{7}$ of the mass of the human body.

These results do not express the total reality. The wastes noted above coming from fossil fuels occur naturally in the biosphere and lithosphere. Carbon dioxide, sulphur dioxide, and ash are not foreign materials to the terrestrial biosphere like the radioactive wastes. This does not mean that Man encounters no radioactivity in this normal environment. Also, the "non-dangerous" carbon dioxide from fossil fuels can have a major impact on the environment.

6.10.2 Radioactive waste from nuclear energy

Radioactivity is a special kind of energy source. Radioactivity absorbed by the human body is measured by the amount of energy released. A rather high dose of ionising radiation, which causes very strong changes when absorbed in the human body, including death in half the cases, equals only 5 grays (Gr). One gray is equivalent to 1 J/kg (1 Gray = 100 rad). For an adult weighing 70 kg this lethal dose is:

$$5 \, \text{Gr} \times 70 \, \text{kg} = 350 \, \text{J}.$$

The same amount of energy if absorbed as heat will raise the body temperature by:

$$\Delta T = \frac{350 \, \text{J}}{(4.1 \, \text{J/g} \cdot \text{K}) \times (70{,}000 \, \text{g/body})} \cong 0.001 \, \text{K},$$

that is, one-thousandth of a degree. This makes it clear that material nonradioactive wastes and radioactive wastes cannot be treated in the same way.

Radioactivity is not a strange occurrence in the natural environment. Figure 6.21 shows the important radioactive sources in the terrestrial environment, both natural and artificial.

The total absorbed dose of natural radioactivity equals approx. 0.1 Gr for the total lifetime of an individual. Based on this information it is now possible to discuss the impact of pollutants from various types of power plants with a reference output of 1 GW (el).

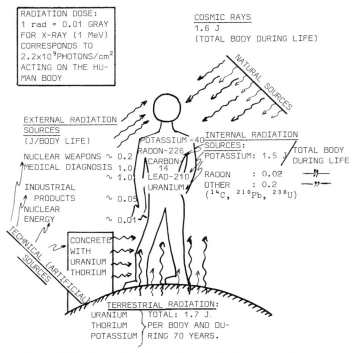

Figure 6.21. Natural radioactivity in the human environment.

6.10.3 Are fission reactors really dangerous?

Fission energy raises some dangerous and complicated problems which have to be solved. Radioactive waste (fission products) are only one of them (see Section 6.10.4); others are:

- the use of uranium-233, uranium-235, and plutonium-239 in nuclear weapons;
- the risk (very small but not negligible) of a catastrophic event in a power reactor resulting in the release of highly toxic radioactive materials into the environment;
- the long-lived radioactive nuclides contained in the reactor's structural materials will be a continuing burden for the biosphere.

Concerning the first problem, it has to be stressed that the production of nuclear explosives does not have to be coupled with the existence of nuclear power stations. Uranium-235 exists in nature and can be isotopically separated from natural uranium by centrifugation, diffusion, or ultrasonic nozzles. Similarly, Pu-239 can be obtained independently of any power reactor via special reactors or accelerators working closer to ambient temperatures and at ambient pressures. The problem of a catastrophic

6.10 Energy Production as a Source of Dangerous Waste and Environmental Problems

accident in a power plant seems to be positively solved, in that the present experience with hundreds of reactors operating for tens of years demonstrates the high degree of safety attained to date.

6.10.4 Radioactive waste and its management

Even when working perfectly and without disturbance, each reactor is a source of a large amount of radioactive waste.

A typical reactor with an electrical power of 1 GW produces approx. 1.8×10^{20} radioactive atoms every second. After a long period of time, having reached a stable state, the radioactivity of the fission products reaches approx. 1×10^{20} disintegrations/sec, in the old units 3 giga-curies (3×10^9 Ci), or in the new units 1×10^{20} bequerels (1×10^{20} Bq). Most of the fission products can be separated out during reprocessing of the irradiated fuel. A well-designed reprocessing plant allows these radioactive wastes to be formed into solidified nonsoluble material.

The amount of radioactive wastes is relatively small. For the 1 GW electrical power reactor the annual production of highly radioactive waste equals only 1000 kg. But this material is highly dangerous. What does one do when faced with a dangerous substance which emits invisible but potentially lethal radiation, and continues to do so for hundreds of thousands of years?

A very important question, and the answer is not so easy to give: Is radioactivity really so dangerous that all contact with it causes irreversible damage? The discussion will concentrate on only one source, a very important one, of natural radioactivity: the Earth's crust.

The average amount of uranium in the Earth's crust equals 3–4 g per 1000 kg of rock. The amount of thorium is four times larger. However we consider here only the uranium.

In a highly populated area with 200 inhabitants per square kilometre, a surface of 5000 m² is available to each inhabitant. The Earth's crust down to a depth of 10 m lying under the 5000 m² contains 600 kg of uranium. This amount of uranium in the natural state corresponds to a radioactivity of approx. 10^{11} disintegration per second, that is, approx. 100 GBq (old units: 3 Ci).

Quite frequently, even in areas with a high population density, uranium ores can be found having 0.2% uranium. In the case where the ore bed is 0.5 m thick, the amount of uranium under an area of 5000 m² is 10,000 kg, giving a much higher irradiation level (Figure 6.21). In many countries densely populated areas lie over massive layers of granite. Taking a layer of granite only 10 m thick, the amount of uranium in a 5000 m² area is 4500 kg.

This data helps to indicate ways in which the radioactive waste problem of the power reactor might be managed.

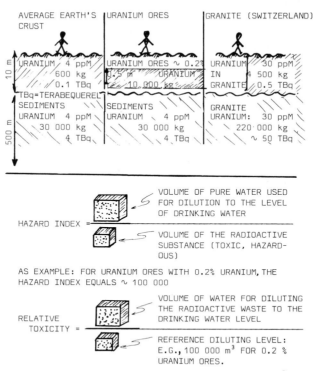

Figure 6.22. The natural radioactivity of the rocks under each 5000 m² "owned" by the individual.

As a reference measure of natural radioactivity, uranium ores of 0.2% can be used. As an absolute hazard index, we can use the volume of pure water needed to dilute a volume of uranium ores to a safe drinking level. It is found that 1 m³ of uranium ore with 0.2% uranium must be diluted by 100,000 m³ of pure water to reach a safe level for drinking water (Figure 6.22).

From this, a relative hazard index can be formulated as being equal to the ratio of the absolute hazard index of a given substance to the absolute hazard index of the 0.2% natural uranium ore. This number is called the relative toxicity. Figure 6.22 gives the relative toxicity of the most important fission products.

From Figure 6.23 it can be seen that the two dominant fission products strontium-90 and caesium-137 have a relative toxicity after 600 years equal to the uranium ores with 0.2% uranium. After that point the main contributor becomes iodine-129, which is a long-lived radionuclide. The actinides, plutonium, americium, curium, and californium, also remain important contributors to the overall toxicity of fission products for a long period.

A method exists for destroying some of the dangerous radioactive nuclides – it is called "transmutation". Intensive bombardment by neutrons results in

6.10 Energy Production as a Source of Dangerous Waste and Environmental Problems 185

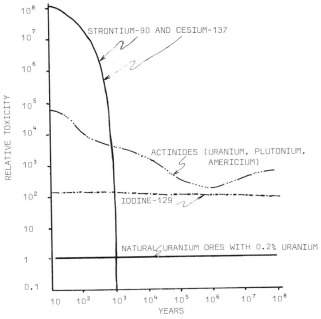

Figure 6.23. Radioactive wastes are very long-lived.

nuclear reactions, the products of which are often short-lived nuclides which decay in some hours or days to stable nonradioactive substances. The actinides are particularly good candidates for this transmutation.

The long-lived fission products can be mixed into ceramics, glass, etc., and then enclosed in steel containers and stored in water-tight rocks, for example, rock salt, granite, and anhydride.

The eventual corrosion or degradation of these containers will allow the release of the dangerous substances into the rocks, but the whole pathway back to the biosphere can be made long enough to ensure that, by the time it reaches the surface and man, the radioactivity will have fallen to negligible levels.

The proposed method of storage underground for a thousand years or longer thus appears to achieve the aim of shielding mankind from the dangerous products of its own civilisation.

6.10.5 Fusion: The controlled thermonuclear reactor – Is this the "clean" solution?

There is a popular opinion that the controlled thermonuclear (fusion) reactors – that is, the reactors of the future burning deuterium and tritium – will be much safer than the fission reactors. In reality this is not so certain. Here are the reasons:

- The fusion reactors contain tritium, a rather dangerous radioactive nuclide.
- Tritium and deuterium, being fuel for fusion reactors, are also the explosive agents for thermonuclear bombs.
- Fusion reactors are very complex machines, using extremely advanced technologies, and experience with some of its properties and possibilities – very high temperatures (up to 100 million K), very intense magnetic fields, very low temperatures for superconductors, and possibly laser technology – is limited.
- The present conceptions of such reactors would produce large amounts of neutrons with a very high energy (14 MeV) and high intensity (10^{14} n/cm$^2 \cdot$ s). This results in the irradiation of the structural materials, which are transformed into radioactive wastes, and which are also dangerous and not entirely short-lived.
- The neutron flux of fusion reactors can be relatively easily used for transforming natural uranium into plutonium with all the implications for proliferation of nuclear weapons (the so-called hybrid-fusion reactors).

6.10.6 Thermal waste, the local and global problem

Every form of energy on this planet sooner or later must be transformed into thermal energy, with a temperature somewhat higher than that of the environment and, finally, transported by infrared radiation into cold cosmic space.

There are at least two different problems, both of which are of the highest importance for the continuing evolution of man's civilisation:

- the total global heat production, whether from fossil chemical fuels or from fissile or fusion fuels, which has to be set against the energy balance of the Earth ($\sim 120{,}000$ TW);
- the local energy flux, coming from the various energy transformation processes, for example, cooling towers of modern power stations and heat from city centres, which has to be related to the mean heat flux on the Earth's surface (equivalent to approx. 170 W/m^2 (Figure 6.24)).

Both problems have to be examined in terms of their impact on the environment, but even today it seems clear that, allowing a reasonable evolutionary path for mankind, both problems are solvable.

6.10.7 Surface waste in the production of energy

The following data seem to be obvious and show the impact of power technologies on the most vital and really limited resource on this planet, its surface.

6.10 Energy Production as a Source of Dangerous Waste and Environmental Problems   187

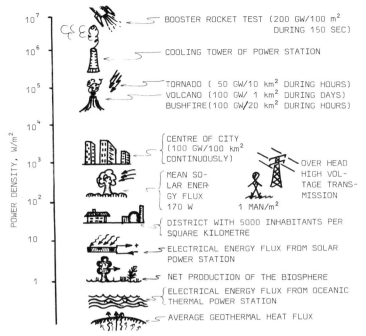

Figure 6.24. The local energy flux density.

These are the values for specific power density for the primary carriers of energy:

- energy yield of uranium in uranium-containing ores at 0.1 % concentration with a thickness of 5 m after full conversion at 50 % efficiency is 50,000 PJ/km² · year;
- energy yield of average oil field ≈ 5 PJ/km² · year;
- energy yield of surface-mined coal in the USA = $\sim$ 100–200 PJ/km² · year;
- energy yield (total) over a whole year from the Sun at an average flux of 168 W/m², which corresponds to 5.3 PJ/km² · year;
- energy from the biomass at 2 % efficiency (very optimistic) using solar energy, which corresponds to 0.10 PJ/km² · year;
- energy of geothermal sources, taking an average for the continents, which corresponds to 0.004 PJ/km² · year.

Of course, only the solar and geothermal energies are fully renewable, the others being "once-through" systems.

For the sake of illustration, one modern man (8 kW per capita) uses 0.00025 PJ per year. With a population density of 200/km² the total requirement is 0.05 PJ/km² · year. No further comment is required.

6.11 The Economics of Energy Production

6.11.1 The energy cost of energy

The primary sources of energy such as solar light, heat of the Earth's crust, coal or oil deep in sedimentary rocks, and uranium ores or deuterium in water, are there "for free". They can be had without payment (neglecting the impact of the local monopoly on price). But to transform these sources into forms of energy suitable for direct use, a rather high price must be paid, a price expressed in energy units.

A simple example can illustrate the problem. Let uranium be extracted from granite, which contains 50 ppM uranium and thorium. In 210,000 m³ (a cube with 60-m sides) of granite one finds 26,000 kg of fissionable material. After total fission in a breeder system 2250×10^{15} J are produced, that is, 2250 PJ (Figure 6.25). With an efficiency of 0.33 this can be transformed into an electrical energy of 750 PJ (el). If this energy is produced over 30 years, the power station has a capacity of 1 GW (el).

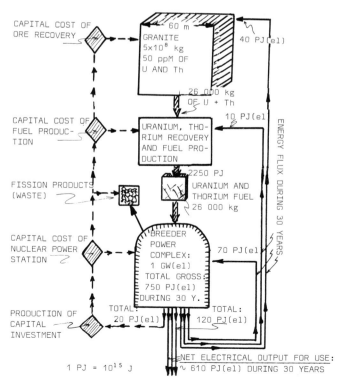

Figure 6.25. The energy cost of energy production.

6.11 The Economics of Energy Production

The construction requires a plant for extracting the uranium from granite, a plant for purifying the uranium (and thorium), and the components of the nuclear power system, which include:

- fabrication of fuel elements,
- construction of power station,
- reprocessing of irradiated fuel, and
- waste management and storage.

All these industrial objects need approx. 20 PJ of electrical energy for construction.

Approximately 120 PJ (el) are used for the exploitation of this source over 30 years. In the same period, the power station of 1 GW (el) produces:

$$1 \times 10^9 \text{ W (el)} \times (3.15 \times 10^7 \text{ s/year}) \times (0.8) \times 30 \text{ y} = 750 \times 10^{15} \text{ J} = 750 \text{ PJ}$$

$$\frac{\text{Gross production}}{750 \text{ PJ}} - \frac{\text{construction}}{20 \text{ PJ}} - \frac{\text{exploitation}}{120 \text{ PJ}} = \frac{\text{net production}}{610 \text{ PJ (el)}}.$$

The net output equals: 610 PJ (el)/750 PJ (el) = 80%.

Even the oil industry, based on a good natural resource, has a net output lower than 70% when all energy inputs have been calculated:

- the geological search for oil reserves,
- drilling, capital charges and operating costs,
- oil pipelines, capital and operating costs,
- tankers for transport, capital and operating costs,
- storage and distribution costs, and
- losses in the refineries.

One important question is: How large will the "energy costs" be for the new energy sources, fusion and solar power stations?

6.11.2 What is the price of energy?

The different forms of energy have not only different levels of quality – for example, high-level electrical energy and low-quality heat for space heating – but also different properties and different roles to play in Man's civilisation. All of these result in different prices for the various forms.

The price of 1 GJ of energy expressed in US dollars (1982) varies:

- 0.9 for coal at the pit head,
- 4.7 for crude oil at the consumer,
- 8.5 for domestic electricity,
- 500 for human food,
- 20,000 for average human labour, and
- 200,000 for intellectual labour (Figure 6.26).

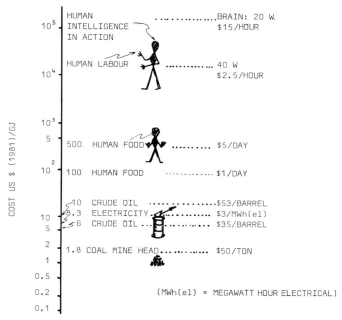

Figure 6.26. The cost of energy, human and technological.

A man in an industrialised country uses approx. 200 times more technological energy then his own human labour:

$$\frac{\text{man: 40 watts}}{\text{technological energy demand 8000 W/capita}} = \frac{1}{200}.$$

With an average price for technological energy of $6/GJ, human labour should cost only $6 \times 200 = \$1200/GJ$. The price of $2.5/ hour of labour, which corresponds to $20,000/GJ, is much higher than calculated from these simple ideas. Human intelligence in action is the highest, most expensive, and most valuable form of energy.

6.12 What Are the Conclusions for the Future Development of Mankind?

From all the questions discussed in this chapter:
- the sources of the primary energy on Earth;
- the nature of the flux and the conversion of solar energy;
- the coupling between solar energy and the climate;
- the energy needs of Man;

6.12 What Are the Conclusions for the Future Development of Mankind?

- the technologies of the use of solar energy in the form of light and fossil fuels;
- the technologies of the use of nuclear energy and resulting environmental problems;
- the energy cost of energy production and the problems of the energy sink which can be the limiting element,

the following conclusions can be drawn for the decisive problem of this book, the future development of mankind:

- The primary energy flow of free energy, solar light, is assured for billions of years.
- The energy needs of human civilisation can be met by means of solar power technologies or by nuclear power technology – this is, fission – and probably also in the future: fusion.
- An increase of energy consumption per capita is necessary if all the needs of human civilisation are to be fulfilled.
- The negative impact of the increasing energy flux can be solved in the future.

It is trivial to say that these are only the physical bases for the further evolution of mankind. The ultimate choice of the future way lies clearly with Man himself.

CHAPTER 7

The Biosphere: The Coupling of Matter and the Flow of Free Energy

The general struggle for survival of the living creatures is
- not a struggle over primary materials,
- also not a struggle over energy (which in the form of heat is abundantly available in everybody, but unfortunately not convertible to other energy forms),
- but a struggle over entropy which is made available in the transfer of energy from the hot Sun to the cold Earth.

freely translated from
L. Boltzmann
(1844–1906)

7.1 The Biosphere: The Coupling of Matter and the Flow of Free Energy

The evolution of the Universe, the galaxies, the stars, and the planets and other small cold objects is the result of the intimate relationship between matter and energy.

The mean energy flux per unit of mass for the stars of the Main Sequence (see Section 3.3.4) is about 1 mW/kg. Is this large or small? Intuition would say that the energy flux from the stars must be very big. In reality, this belief is far from the truth.

The flow of energy in the terrestrial climasphere (see Section 4.8) is given as:

$$\frac{125{,}000 \times 10^{12}\,\text{W}}{5 \times 10^{18}\,\text{kg/(atmosphere)} + 1 \times 10^{19}\,\text{kg/(30-m layer of ocean)}} = 0.01\,\text{W/kg}.$$

The energy flux in the terrestrial climasphere is thus 10 times higher than that of the Main Sequence stars, for example, the Sun.

It will be shown that in the biosphere the energy flux per unit mass reaches 0.07 W/kg, that is, four times higher than that of the climasphere. In the animal kingdom this factor rises still higher, to 1 W/kg or more. The human brain needs an energy flux of ~ 20 W/kg.

At first glance this result is surprising, but it is nevertheless true that the flow of energy per unit mass for the living being is extremely high and is in the form of energy of the highest order, free energy. This fact further indicates that to support life, large amounts of high-quality energy are needed.

Life is the product of the intimate coupling of the fluxes of matter, free energy, and information.

7.2. The Terrestrial Biosphere: Mass and Productivity

7.2.1 The greatest component of the biosphere is, in terms of mass, in the form of trees

One of the most important questions is: What is the mass of the terrestrial biosphere at the present time?

Most students of this topic agree that, given the present state of our knowledge, the best value of the terestrial biomass converted to mass of dry material is (Figure 7.1):

$$1840 \times 10^{12} \text{ kg} \quad \text{or} \quad 1840 \text{ Pg}.$$

(In this chapter, all mass will be given in units of petagrams, which is also the mass of one cubic kilometre of water.)

To get a feel for this figure, it can be converted to the mean amount of living matter per unit surface area of the Earth:

$$\frac{1840 \times 10^{15} \text{ g}}{5.1 \times 10^{14} \text{ m}^2} = 3.6 \text{ kg/m}^2 \text{ (dry biomass)}.$$

We can assume that the "wet" living material is some 2.3 times greater, therefore the mean amount of living matter on the Earth per unit area is 8.2 kg/m². Taking a mean density for living material of ~ 1 Mg/m³ gives a layer of living matter some 82 mm thick, spread uniformly over the globe. Of course, the layer is not uniform; over the tropics it is much greater (~ 2 orders of magnitude) than at the poles.

Before continuing this discussion, the form of this biomass must be clarified. The relationship between animals and vegetation is, of course, commonly known.

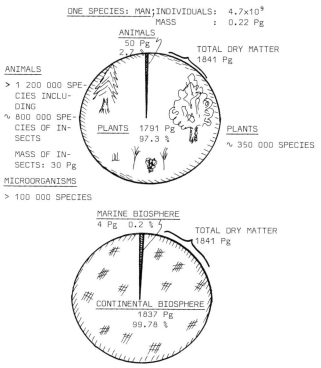

Figure 7.1. On the continents the biomass consists almost entirely of plants. 1 kg of dry biomass contains 0.45 kg of elemental carbon; is equivalent to 2.3 kg net biomass (that is, including bonding water); corresponds to 16.7 MJ of chemical energy; includes also 0.23 kg of "minerals".

However, here intuition can again deceive. More than 97% of the biomass exists in the form of plants, with animals representing no more than 2.7% of the total mass. From a mass point of view the greatest part of the biopopulation is in the form of trees! (See Figure 7.1.)

A similar unexpected picture is obtained if the distribution of the biomass over the continents and oceans is considered. Only 0.21% of the dry biomass is found in the oceans, in spite of the fact that they cover 2/3 of the Earth's surface. The continents contain 99.79% of the biomass. In the present epoch, life is undoubtedly a continental overland phenomenon.

All the foregoing information is valid for the Earth taken as a whole without considering the dramatic regional variations. Figure 7.2 shows the distribution of the biomass on the planet. More details are given in Table 7.1. In the tropical regions the dry biomass can be equivalent to a surface loading of 200 kg/m^2, whereas in the polar regions it falls to 3 kg/m^2 and further to 0.01 kg/m^2 in the oceans (mean value for all oceans).

7.2 The Terrestrial Biosphere: Mass and Productivity

Table 7.1. Biomass, Dry Matter, and Net Productivity

Areas	Surface		Dry Biomass		Net productivity	
	Tm²	%	Pg	%	Pg/year	%
Polar	8.05	1.6	10.5	0.57	0.5	0.3
Boreal	23.2	4.5	335.0	18.2	10.3	6.0
Sub-boreal	22.5	4.5	213.5	11.6	10.3	6.0
Subtropical	24.2	4.8	248.5	13.5	22.4	13.0
Tropical	55.8	10.8	10.29.0	55.9	73.0	42.3
"Land" without glaciers, lakes, subtotal	133.4	26.2	1836.8	99.77	116.5	67.6
Glaciers	13.9	2.7	0	0	0	0
Lakes	2.0	0.4	0.2	0.1	0.7	0.4
Continents, total	149.3	29.3	1837.0	99.78	117.2 [R]	68.0
Oceans, total	361.0	70.7	4.0	0.22	55.3	32.0
Earth, total	510.3	100.0	1841.0	100.0	172.5 [R]	100.0
Continents	149.3	100%	1837.0	100%	117.2	100%
Tropical rain forest [a]	17.0	11.4	755.2	41.1	37.1	31.6
Tropical season forest	7.5	5.0	266.5	14.5	12.4	10.6
Temperate forest, evergreen	5.0	3.3	177.7	9.7	7.0	6.0
Temperate forest, deciduous	7.0	4.7	211.0	11.5	9.2	7.8
Boreal forest	12.0	8.0	339.9	18.5	10.4	8.9
All forest	48.3	32.2	1650.0	89.8	76.1	64.9
Woodland savanna	23.0	15.4	108.0	5.9	16.7	14.2
Grassland	9.0	6.0	14.0	0.8	4.8	4.1
Tundra, alpine	8.0	5.3	5.3	0.3	1.2	1.0
Desert, semidesert	18.0	12.0	12.0	0.6	1.5	1.3
Extreme desert, ice	24.0	16.0	0.4	~0.0	0.1	0.1
Cultivated land [b]	14.0	9.3	15.5	0.8	10.0	8.5
Swamps, marshes	2.0	1.3	30.2	1.6	5.3	4.5
Lakes, streams	2.5	1.7	~0.0	0.0	1.4	1.2
Continents	149.3	100.0	1837.0	100.0	117.2	100.0

[a] A hundred years ago the evergreen tropical lowland forests occupied an area about twice the size of Europe, $\simeq 20\,\text{Tm}^2$. Now they have been reduced to one the size of Europe, $\simeq 10\,\text{Tm}^2$. Each year an area equivalent to that of Great Britain ($0.1\,\text{Tm}^2$) is cut away or degraded.

[b] At present: $17\,\text{Tm}^2$

[R] *Remark*: according to Budyko, the productivity:
Continents: 141 Pg/y;
Earth: 196 Pg/y.

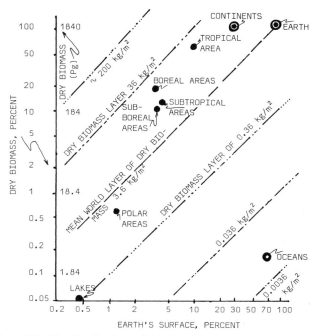

Figure 7.2. The distribution of the dry biomass around the world.

7.2.2 The biosphere's productivity does not match its pattern of distribution

The biomass is a system in steady state. Many of the organisms die in a given time and a similar quantity are born and begin to grow.

Figure 7.3 shows the principal flow of matter in the biosphere. The annual production of the total biosphere is approximately 172.5 Pg. This annual net rate of production is divided between the continental production of 117.2 Pg/year, that is, 68% of the total, and 55.3 Pg/year from the oceans, or 32% of the total (see Figure 7.4). The productivity of different continental regions is shown in Figure 7.5.

It can be seen from Figure 7.5 that the forest is the most important producer of the continental biomass, approx. 73.9 Pg/year or 43% of the total global biomass production.

It is possible to calculate the amount of solar energy which has been transformed into the chemical energy of dry biomass. From the previous data (Figure 7.1), we know that 1 kg of dry biomass contains, on the average, approx. 16.7 MJ of bonded chemical energy.

A productivity of $1 \text{ kg/m}^2 \cdot \text{year}$ corresponds to an energy flux of:

$$\frac{1 \text{ kg dry biomass}}{\text{m}^2 \cdot \text{year}} = \frac{16.7 \text{ MJ}}{\text{m}^2 \cdot \text{year}} = \frac{0.53 \text{ J}}{\text{m}^2 \cdot \text{s}} = 0.53 \frac{\text{W}}{\text{m}^2}.$$

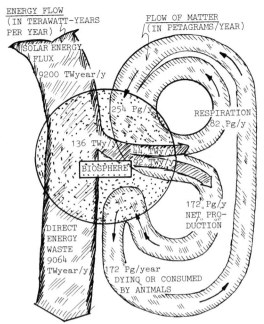

Figure 7.3. The biosphere is a steady-state system.
1 TW year $= 10^{12}$ W $\times (3.15 \times 10^{7}$ s/year$) = 3.15 \times 10^{19}$ J.

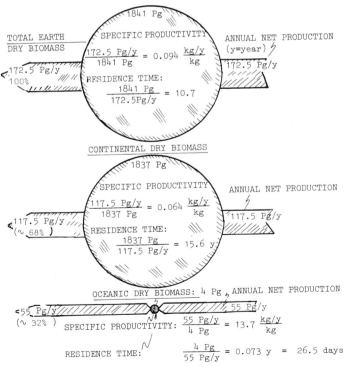

Figure 7.4. Comparing the continental and marine biospheres.

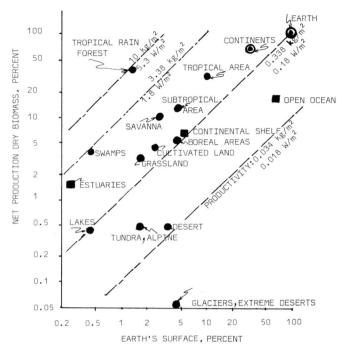

Figure 7.5. Net productivity in the different regions and climatic zones.

Figure 7.5 parallels the annual productivity in kilograms of biomass per m², the corresponding energy flux in W/m².

In the most productive areas, such as the tropical rain forests, the biomass production rises to 10 kg/m² year, equivalent to an energy flux of 5 W/m². In deserts the productivity falls to 0.04 kg/m · year, or only 0.02 W/m². The ratio of these extremes is 250:1.

7.2.3 The surprisingly simple chemical composition of the biosphere

The living organism is the most highly ordered material in the world. Further, the functions, behaviour, and patterns of life are many-sided and varied. Nevertheless, the chemical composition of these living creatures is quite simple.

Half of the atoms are hydrogen (Figure 7.6), the most abundant element in the Universe, also the most abundant in the biosphere. The next elements on the scale of abundance in living matter are oxygen and carbon which together make up almost all the remaining atoms. Additional elements total not more than about 1 %, nitrogen being the largest part at 0.27 %. Even such "important" elements as phosphorus and sulphur amount to only ∼300 atoms per million.

7.3 The Magnitude of the Flow of Energy in the Biomass

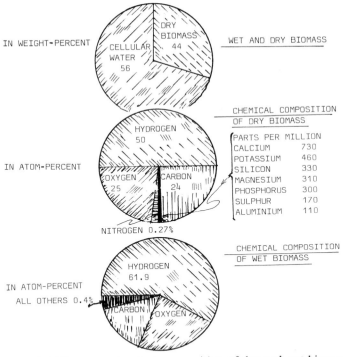

Figure 7.6. The mean chemical composition of dry and wet biomass.

All this is valid for the dry biomass. Since intracellular water plays a very important role in the living processes, the true composition of the living matter must be corrected to give a very large increase in the amount of hydrogen and oxygen actually present.

7.3 The Magnitude of the Flow of Energy in the Biomass

7.3.1 The direct net flux of energy in the biosphere is some 92 TW

The total net productivity of the biosphere has been estimated as 172.5 Pg/year.

One gram of dry biomass undergoing photosynthesis consumes about 16.7 kJ (see Figure 7.1). From this we can calculate the annual amount of solar energy used in photosynthesis:

$$(172.5 \times 10^{15} \text{ g/year}) \times (16.7 \times 10^3 \text{ J/g}) = 2.88 \times 10^{21} \text{ J/year},$$

which corresponds to an energy flux of:

$$\frac{2.88 \times 10^{21} \text{ J/year}}{3.15 \times 10^7 \text{ s/year}} = 9.15 \times 10^{13} \text{ J/s} \simeq 9.2 \times 10^{13} \text{ W} = 92 \text{ TW}.$$

In this calculation only the net productivity has been taken into account. The gross productivity comes to 255 Pg/year, approx. 1.48 times more than the net value. Thus the total solar energy flux required for photosynthesis is 136 TW (see also Figure 7.3).

Is it then accurate to say that this 136 TW, which represents only a fraction of the total solar flux, is sufficient for supporting life on this planet? Is only a small fraction of the available free solar energy required by the biosphere?

The answer is emphatically no: The amount of solar energy needed for the steady state of the biosphere is very much larger than the value given by the simplified calculation above. The total demand for energy from the biosphere is some orders of magnitude greater than 136 TW.

7.3.2 The total solar energy flux consumed by the biosphere

The reasons why the efficiency of photosynthesis is low are rather easy to state and can be summarised as follows (Figure 7.7):

– Green plants cover the surface of the Earth only at certain seasons. At other times the leaves are either too old, too young, or absent.

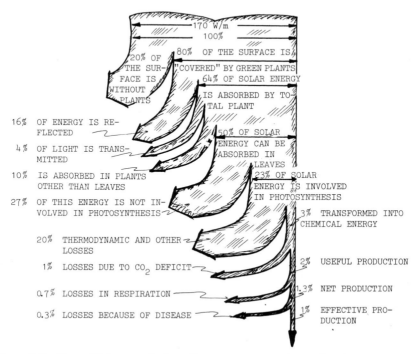

Figure 7.7. The efficiency of the direct net conversion of solar energy via photosynthesis.

7.3 The Magnitude of the Flow of Energy in the Biomass

- The mean solar energy flux is estimated at no more than 170 W/m^2, from a total of 340 W/m^2 (see Section 6.2.1).
- A part of the solar radiation is reflected back by the leaves.
- The leaves are partially transparent.
- Part of the sunlight is absorbed by other parts of the plants.
- Less than half the Sun's spectrum is able to promote photosynthesis; the rest has too little specific energy (less than 1 eV/photon).
- The thermodynamic efficiency of the photosynthesis process equals 13% which, in fact, is rather surprisingly high.
- About 1/4–1/3 of the newly synthesised biomass is used by the plant internally for "food", for life processes, and for respiration.
- Some of the biomass is destroyed by diseases.

Taking all these factors together, only 1% of the total solar energy can be converted to the net biomass produced. From this, it must be clear that the real need of the biosphere for solar energy for photosynthesis is at least 100 times greater than the direct net energy used in the process, itself some 92 TW. Thus the total solar energy involved in photosynthesis is at least 9200 TW, but even this figure is not complete (see Section 7.3.4).

7.3.3 The biosphere in the past

It can be estimated that the total dry biomass produced during the history of the Earth has been between 1×10^{22} and 1×10^{28} kg. There can be little doubt that the numbers of individual animals and plants and their total biomass are greater now than they were in the Cambrian (~ 500 million years ago). Very likely they are also greater than they have been at most times since the beginning of life. Even if we include microorganisms (the number of which on Earth today is not known with any reasonable confidence), it is probable that the number of living individuals and the number of species has increased.

It can be postulated with rather high probability that 4.5 billion years ago the Sun was 35% less luminous than it is now. The increase of solar luminosity can be assumed to be a linear increase over the Earth's history. This assumption is significant because it can result in major changes in the global climate. But how big was the biosphere in the not-so-distant past? Is it really true that in Europe today only 25% of the surface is covered by forest, while about 2500 years ago the forest covered 90%? And the same in India, China, North Africa, North America? Is it not true that, in the thousand years since the contraction of the north continental ice shield in Siberia and Canada, the tundra area has increased?

One of the highest accumulation rates of carbon must have been in the soils and forests of the boreal and tundra zones. Since the retreat of glacial ice 5000 to 15,000 years ago, these regions accumulated about 5×10^{14} kg of carbon. The present total of dry biomass equals 1841 Pg (1.841×10^{15} kg) (see

Fig. 7.1), which corresponds to 8.3×10^{14} kg of carbon. The total accumulation in tundra soils is of the same order of magnitude.

It has been assumed that in the last two decades the tropical forests have covered ~ 16 million km^2 of the continents. Recent calculations give a surface of only ~ 10 million km^2 covered by the tropical forests. It seems that each year approx. 0.1 million km^2 of tropical forest are destroyed. At this rate, in ~ 100 years we could lose the total area of tropical forests, and we must remember that these forests represent a very significant proportion of the total global biomass.

In recent times it has been calculated that tree felling in the world, especially in the tropical rain forests (e.g. Amazonia) may reach 10^{12} kg/year expressed as carbon. Although it is not yet possible to be certain, the indications suggest that the higher figure is closer to the truth and even this figure may turn out to be too low.

The best estimation of the annually burned biomass equals $\sim 5.8 \times 10^{12}$ kg expressed as carbon, that is, approx. 1.3×10^{13} kg of dry biomass. This corresponds to $\sim 10\%$ of the annual net production of the biosphere!

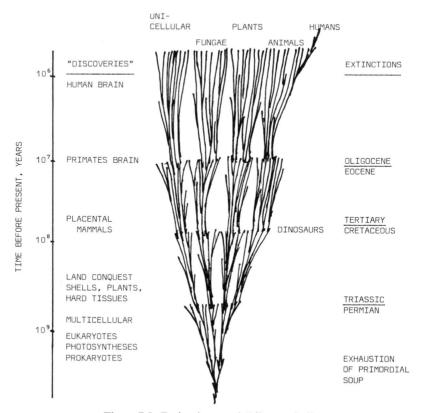

Figure 7.8. Extinctions and "discoveries".

7.3 The Magnitude of the Flow of Energy in the Biomass

The present rate of the destruction of the biosphere is doubtless a very large phenomenon. A solution for this threat to the terrestrial biosphere, including mankind, must be found in the next decades.

In the past evolution of the biosphere, some very difficult periods called "Great Dyings" are well known (see Figure 7.8). The reasons for these catastrophes are not well known.

Here are some hypothetical causes of the mass extinction of species:

- fall of comets (10 km diameter, 10^{15} kg)
- fall of meteorites
- climatic change due to a large gas-cloud on the galactic plane
- decrease of the ocean level due to climatic changes.

These catastrophes seem to be repeating/recurring each 28 ± 2 million years.

The primordial causes of these recurring falls of comets and/or meteorites are as follows:

- stellar companion of the Sun (brown dwarf with mass of 0.01 solar mass)
- oscillation of Sun perpendicular to the galactic plane.

Figure 7.9 shows some possible scenarios for the mass extinction.

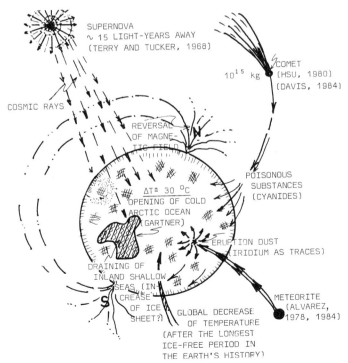

Figure 7.9. Models for the mass extinction.

During its evolution, the biosphere shows not only the impact of the "Great Dying" but also very spectacular "Great Discoveries" and "Great Conquests", including:

- the discovery of photosynthesis,
- the discovery of eukaryotic cells (organisms),
- the discovery of multicellular organisms,
- the discovery of hard tissues,
- the conquest of the land,
- the discovery of the central neurvous system,
- the discovery of intelligence.

7.3.4 The green plant is not only a synthesiser, it is also a water vapouriser

That factory in miniature, the green plant, requires a large supply of water for at least three reasons (Figure 7.10):

a) In the synthesis of new dry biomaterial, calculated at 3.1 kg/m² per year, an additional 4 kg of water must be included.
b) The plant must cool itself, and about 50% of the absorbed solar energy is used to vapourise water, that is, 85 W/m² (Figure 7.7) which, at an efficiency of 0.25, is equivalent to the vapourisation of 260 kg of water.

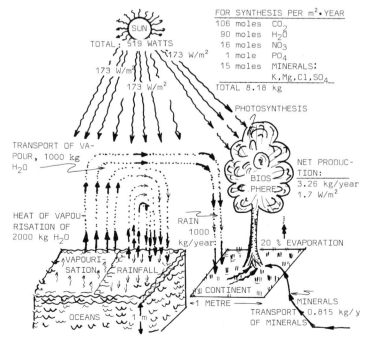

Figure 7.10. The maximum efficiency of the synthesis by green plants.

7.3 The Magnitude of the Flow of Energy in the Biomass

c) This flow of water also acts as a transport medium for 815 grams of minerals needed in the synthesis of 3.1 kg of the dry biomass (note: the mean salinity of this water is 0.31 %, that is, 3.1 g minerals per 1 kg water (Figure 7.10)).

The total amount of energy required for vapourisation of water is thus much greater than that needed purely for photosynthesis. Even this is not the end of the story.

In the model chosen here the green plant vapourises about 260 kg of water per year. For this to be possible, an even larger amount of water, in the form of rain, must fall on the surface. A first approximation gives a rainfall requirement of a layer 1 m thick, that is, 1000 kg of water per square metre per year.

Most of this water is transported from the oceans in the form of rain clouds. To support this transportation a large flux of solar energy must be utilised. The real productivity on the continents is much lower, at 0.785 kg/m^2 year (Table 7.1). This corresponds to an effective energy flux of $\sim 0.42 \text{ W/m}^2$ for the dry material. From the estimate above, the total solar flux required for biosynthesis per square metre of land area and two square metres of ocean (including rain making) is 500 W. The real net efficiency of the biosphere is:

$$\frac{0.42 \text{ W/m}^2}{500 \text{ W/3m}^2} = 0.08 \%.$$

To obtain the continental productivity of 117.2 Pg/year, which corresponds to 61.6 TW directly used in photosynthesis, assuming a total efficiency of 0.08 %, the total solar energy flux needed for the continental biosphere can be given as:

$$\frac{117.2 \text{ Pg/year} \times 16.7 \text{ kJ/g}}{0.0008 \times (3.15 \times 10^7 \text{ s/year})} = 77,000 \text{ TW}.$$

Here is a surprise: To support the continental biosphere having a net annual productivity of 117.5 Pg, an energy flux of 77,000 TW is needed at the surface of the continents and on the oceans.

It must be remembered that the total solar energy flux which reaches the surface equals $\sim 87,000$ TW (Figure 6.3). This is unexpected, yet at the same time not without logic. The whole of the continental biosphere uses the whole of the available solar energy flux, including that of the oceans.

It must be remembered that in all of these estimates the mean annual productivity is only 0.785 kg/m^2. The "theoretically" possible productivity is 3.1 kg/m^2 per year, that is, four times greater. This is a rather important result which we will return to later in this chapter.

7.4 The Biosphere as a Source of Food for Mankind

7.4.1 How much free energy in the form of food does Man need?

Man is a part of the biosphere. He needs a continuous supply of:

- free energy for maintaining his internal order and for carrying out work, and
- structural materials for maintaining his system.

Both needs can only be met by the consumption of some well-defined chemical compounds present in some animals and plants.

Take first the flux of free energy in the food required by man. The mean value of his needs equals:

$$\frac{2500 \text{ kcal}}{\text{day}} = \frac{10.5 \text{ MJ}}{24 \text{ h}} \cong 120 \frac{\text{J}}{\text{s}} \cong 120 \text{ W}.$$

This flow of free energy is used for the following purposes:

Internal metabolism	85 W	
brain		20 W
gastrointestinal tract		20 W
heart		5 W
kidney		8 W
muscles		15 W
other		17 W
Movement (work of muscles)	15 W	
Labour (work of muscles)	20 W	
Total	120 W	

All these values are calculated over a long period of time to derive the mean values given here. For short periods man can increase his activity to much higher levels; for example, during a march with a speed of 6 km/hour and a 20-kg load, the energy flux rises to ∼600 W, that is, 5 times greater than the mean energy flux of 120 W. Under these conditions the energy used in walking is only 200 W, leaving the remaining 400 W for metabolic processes. Such intensive work is possible for only a few hours per day and with a higher than average food intake (see also Figure 6.7).

7.4.2 Man requires numerous structural materials for his body

For each living being the body is in a "stable" state, where the rate of destruction equals the rate of renovation. Some organisms have the ability to synthesise almost all the structural materials they need. This is not true for animals, particularly not for man.

7.4 The Biosphere as a Source of Food for Mankind

Man needs a continuous input of the following materials in well-defined amounts:

- 8 amino acids, the so-called essential amino acids: leucine, isoleucine, lysine, valine, methionine, phenylalanine, threonine, and, for babies, also histidine. The rest of the 12 amino acids needed for the body's structure can be synthesised in the body itself;
- 3 essential fatty acids: arachidonic, linoleic, and linolenic;
- 14 vitamins: B-1, B-2, B-12, C, A, D, E, K, niacine, and others;
- 17 elements: sodium, potassium, magnesium, chlorine, phosphorus, sulphur, iron, fluorine, cobalt, copper, manganese, molybdenum, selenium, zinc, nickel, tin, and aluminium. This does not include the main components: hydrogen, oxygen, carbon and nitrogen;
- water as an internal medium;
- air as a carrier of oxygen.

Figure 7.11 gives the total required input per man per day. From all these one might assume that, since life cannot be maintained without sufficient food, the energy contents of free energy in the food would be well defined. Unfortunately, this is not true – we know more about protein needs than we do about energy (calories) needs. A better definition of energy input needs and how individuals adapt to high or low food energy intakes is required.

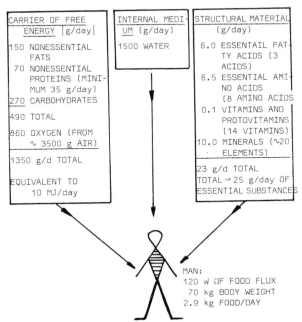

Figure 7.11. The daily influx of energy carriers and structural materials for the average man.

Dietary surveys indicate that the per-capita food consumption in Kerala (India) is near the minimum. In spite of this, the life expectancy is comparable to that in developed nations. Also, the synonymously used terms "protein" and "good nutrition" are not self-evidently equivalent. Even human breast milk is, in fact, a "low-protein food". The total content is approx. 5–6 % of the total "calories".

7.4.3 The winning of food from the biosphere

Ten thousand years ago, Man the hunter/gatherer consumed approximately 100 different edible plants, roots, and fruits and about 50 different types of flesh, mostly having little fat. This was sufficient to give him all he needed in the way of energy, structural materials, and essential trace elements.

Sufficient food of this type can be produced on this planet for a human population of only some 20 million. However, since the last ice age, that is, 10,000 years ago (or about 500 human generations), a dramatic change in the situation has occurred.

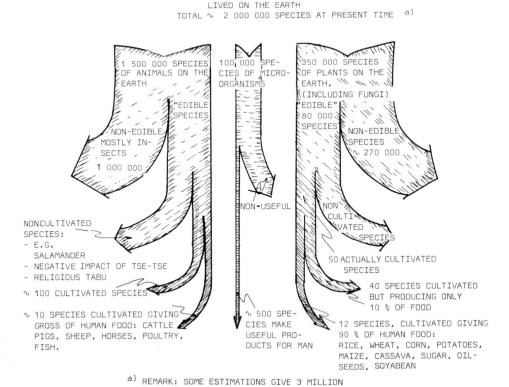

Figure 7.12. Plants and animals used by man as sources of nutrition.

Instead of the 100 different edible plants man now consumes mostly nine vegetables or vegetable products very rich in starch (carbohydrates) such as rice, wheat, maize, and potatoes. Instead of the 50 low-fat meats man today consumes no more than three types of meat with high-fat content (Figure 7.12). This major change arises from the need to increase the productivity of the plants and animals consumed. It is clear that the earlier choice of food had a low productivity. Nevertheless, with the continuing development of mankind, supplying food will become one of the most important and intractable problems.

The total food and feed production (cereals, crops, hay, and grass) recalculated in grain-equivalents equals $\sim 3.7 \times 10^{12}$ kg/year. This corresponds to approx. 880 kg grain-equivalents per capita during each year. Because 1 kg grain-equivalent equals 13.9 MJ (Figure 7.1), the total food and feed average production per capita equals 390 W, which is distributed in following way:

– food for humans   105 W
– feed for animals   245 W
– seed, losses   40 W

7.5 Agriculture, Source of Food for Humans

7.5.1 Agricultural requirements of the average man

The world population today has passed the 4.8-billion mark. The total net food directly consumed is

$$(4.8 \times 10^9 \text{ people}) \times (110 \text{ W/capita}) = 0.54 \text{ TW}.$$

Taking the total net energy flux to the biosphere of ~ 92 TW, the net food for mankind is about 0.6%. The real consumption of dry biomass by man is actually some 8–10 times higher, as will be seen later, reaching about 5% of the total energy flux to the biosphere. To illustrate this more clearly, the food products for the average man are analysed in terms of his agricultural needs (see Figure 7.13).

Man needs a semi-continuous input of food to provide 120 W. Assuming a dry food content of 16.5 MJ/kg, the daily per-capita intake should be:

$$\frac{120 \text{ W} \times (8.64 \times 10^4 \text{ s/day})}{16.5 \times 10^6 \text{ J/kg}} = 0.65 \text{ kg dry biomass/day}.$$

This value corresponds approximately to that given in Figure 7.11 (dry food contains ~ 490 g of energy carrier and ~ 25 g structural material, totalling 0.5 kg of dry biomass per day). However, Man cannot consume only vegetables.

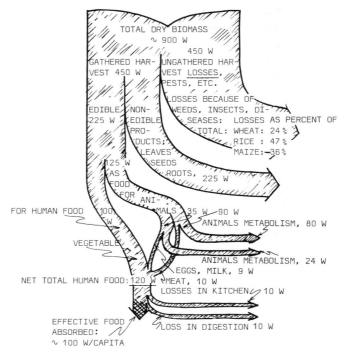

Figure 7.13. Energy flows from the biomass produced by cultivation to provide human food.

The present methods of producing the amounts of food required are shown in Figure 7.13.

From this very simplified breakdown it can be seen that to supply the necessary dry biomass intake for Man, some 900 W are required. This is about 8–9 times more than the effective net intake of food by the human body.

Now comes the question of how large the area devoted to agriculture must be to produce such quantities of food. A simple calculation, assuming the total agricultural efficiency is 0.5 W/m² or 0.95 kg dry biomass per m² per year, of the area then required is:

$$\frac{900 \text{ W}}{0.5 \text{ W/m}^2} = 1800 \text{ m}^2.$$

In reality, the agricultural area per capita today is about 3500 m², or twice the "theoretical" value as calculated above and, in spite of this, mankind is very far from being well fed (Figure 7.14).

Taking the present day state of agriculture, the net efficiency of human food production is given by:

$$\text{net average efficiency} = \frac{\text{food } 110 \text{ W/capita}}{\text{sunshine } 600{,}000 \text{ W}/3500 \text{ m}^2}$$
$$= 2 \times 10^{-4} = 0.02\,\%.$$

7.5 Agriculture, Source of Food for Humans

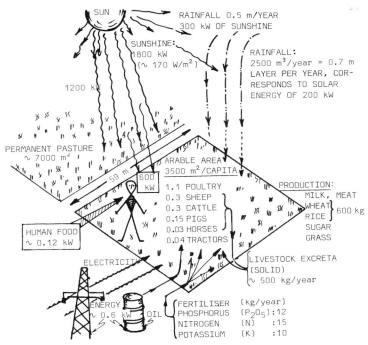

Figure 7.14. The agricultural infrastructure required to feed the average man today, in a developed country.

These are the reasons for the unbelievably small efficiency of agriculture:

- A large part of the cultivated area is not producing for a large part of the year due to lack of water, fertiliser, equipment, seeds, light, or good weather.
- One-tenth of the area is used to cultivate non-food materials, for example, cotton, tobacco, coffee, tea, and construction materials.

So far, mean values for the Earth as a whole have been used.

7.5.2 Human food quality

This aspect is of the highest importance. The quality of food as discussed here is governed by the presence of the following essential components (see Figure 7.11):

- essential amino acids,
- essential fatty acids,
- vitamins and protovitamins.

Clearly, today's problem is the provision of proteins, that is, animal proteins, which alone carry the essential amino acids in the necessary proportions. Unfortunately, the price of producing these proteins is rather high.

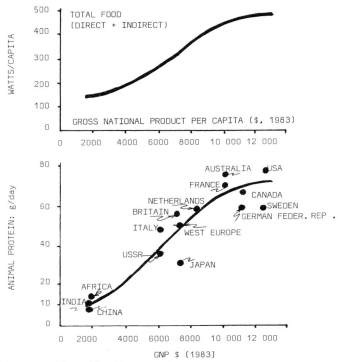

Figure 7.15. Transposition of food increases in step with the per-capita Gross National Product (GNP).

The best types of proteins are the most expensive to produce and consume the most energy in their production. The results are obvious: The rich nations consume an excess of animal protein and are overfed. The poor nations have a deficit of animal proteins and are underfed (Figure 7.15).

The ratio of animal proteins consumed by Americans compared to the Chinese is:
$$\text{ratio } \frac{\text{Americans} \sim 80 \text{ g/d}}{\text{Chinese} \sim 8 \text{ g/d}} = 10.$$

This is the real picture of the under-nutrition of the "third world". The following short table makes this clear.

	Unit	"Rich Man"	"Poor Man"	Ratio "rich/poor"
Food intake, energy	kcal/day	3600	2200	~1.7
	W	175	105	~1.7
Total vegetable consumption, direct and via livestock	W	520	125	4.2
Animal proteins	g/day	80	8	10

Now to the problem of redressing this imbalance. What constraints exist?

7.6 Constraints on the Further Development of Agricultural Production

7.6.1 Can the area under cultivation be increased?

The simplest way to solve the problem of feeding the world more effectively appears to be increasing the area for agriculture: Firstly, how much land is available at the present time (see also Figure 7.16)?

Let us be conservative and assume that at first only the continents are available for the intensive production of food for humans. The problem of the oceans as a food source will be discussed later (see Table 7.2).

Of the total continental area of 149 million km^2, the best-established calculations give us an area suitable for cultivation of about one third of this. The potential agricultural area is thus 45 million km^2 or 45 Tm^2 (Figure 7.16) (1 Tm^2 = 10^8 hectares).

The remainder of the continental area is unsuitable for agriculture for various reasons. Ten percent is always covered by ice, 15% is high mountain region, 15% has an unsuitable climate, and 20% is desert. A further

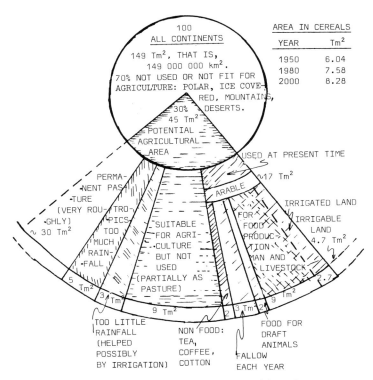

Figure 7.16. Distribution of the world's cultivated areas.

percentage is given over to human habitation, industries, and mines, and only a very small portion is given over to national parks.

Of the 45 Tm², that is ~30%, of continental area available for agriculture, a good part must be considered as unsuitable for the following reasons:

- too much rainfall,
- too little rainfall, and
- too expensive to be transformed into agricultural production.

Referring to the last reason, the limit of what can be transformed may perhaps be set at a minimum cost of $0.2(US)/m². Taking a peasant family of four persons, they need an agricultural area of 4 × 3500 m², or 14,000 m² total. The capital cost to bring such an area into production is $3000(US) or greater. This is likely to be outside the capabilities of the underdeveloped nations. This capital cost is even higher if the peasant family is required to produce food for the non-peasant members of the society. Finally, about 17 Tm² of the Earth's land area appears to be available for agricultural use by mankind, giving an area of 3500 m² per capita (Figure 7.16). Within this 3500 m² are included areas for non-food production (cotton, coffee, etc.), food production for draft animals and reserve land which lies fallow each year.

Summarising, we see that for real food production for mankind only 9 Tm² is used. This can be placed in context as follows:

53% of the total area under cultivation;
20% of the potential agricultural area;
6% of the continental surface;
1.8% of the planet's surface.

It must not be forgotten that the present level of agricultural productivity is at most one third of what is required as a reasonable minimum for a healthy and well-fed mankind. An increase in agricultural production is necessary, broken down as:

10% increase in the energy produced as food is required;
100% increase in the production of animal proteins is necessary.

After allowing for a conversion factor of grain to proteins of 0.15, an increase in gross food (direct human and via livestock) production of about 1.3 times is required.

The real difficulty lies in the future, say in the next two generations. In about 70–80 years, the world population will have risen to 8 billion (8×10^9), twice as high as today's figure. The difficulties this fact presents have been discussed for more than two centuries, since the time of Robert Malthus.

Combining both factors:

- the need to increase the mean level of human consumption by a factor 1.3, and
- the doubling of the world population,

7.6 Constraints on the Further Development of Agricultural Production   215

it can be seen that the world food production must increase by a factor of 2.6 within the next 70–80 years. This corresponds to a yearly increase of ~1.5% over these next 80 years.

Two extreme solutions are possible:

- increase the area under cultivation by 2.6 times, or
- increase the productivity of the present area by 2.6.

Clearly the real solution lies somewhere between the two. In fact, the chance of increasing the agricultural area over the next 70–80 years is limited. At the most, a 40% increase would be a reasonable assumption for the present discussion. This would mean that the present 9 Tm² would increase to ~13 Tm². This is mostly achievable by reducing the area considered as "reserve" from 3 to 2 Tm², reducing the area available to feed draft animals from 2 to 1.5 Tm² (being replaced by mechanical "beasts of burden"), and the recovery of a further 2.5 Tm² now lying fallow.

This total increase of 4 Tm² thus represents a factor of 1.4 increase in production. The rest of the increase, giving a combined factor of 2.6, must come from improvements in productivity. Over the next 70 years an increase of 1.4 would seem possible and corresponds to a yearly increase rate of 0.8%.

In the future, the area per-capita ratio for human food production must decrease from

$$\frac{9 \text{ Tm}^2}{4.5 \times 10^9 \text{ people}} = 2000 \text{ m}^2 \text{ per capita at present}$$

to

$$\frac{13 \text{ Tm}^2}{8 \times 10^9 \text{ people}} = 1600 \text{ m}^2 \text{ per capita in the future.}$$

It can therefore be seen that the problem will be solved not only by increasing the area under cultivation but by a real increase in productivity.

If the above assumptions are correct, then to reach the factor 2.6 increase required, the increase in production must more than double in the next 50 years. This represents a yearly rate of increase of ~1.5%. Is this attainable?

7.6.2 Can agricultural production be doubled over the next 50 years?

This question is one of the most important now facing mankind. It is useful to recall some basic factors:

- The average effective solar energy flux equals ~170 W/m².
- The mean productivity of the continental biosphere equals 0.785 kg of dry biomass per year per m² (Table 7.1). This corresponds to 13 MJ/m² · year or a production "flux" of food of 0.41 W/m², giving a solar conversion efficiency of 0.24% (Figure 7.17).

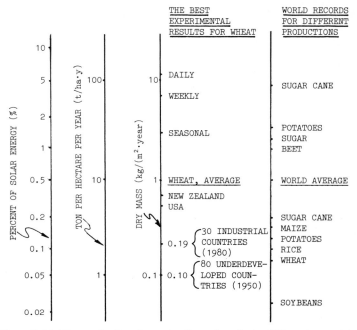

Figure 7.17. The agricultural production of grain in different countries.

The difference between the present mean global grain production and what seems to be desirable in the future is significant but not dramatic.

During the years 1950–1970 the world's grain production rose at an average rate of 2.8 %/year, that is, faster than the population. For rice and wheat in India, Pakistan, and Indonesia, during the years 1965–1975 the rate of production increased by about 5 % annually.

The discussion above requires an increase in annual grain production from 0.2 to 0.3 kg/m² over 50 years and thus an annual rate of increase of only 1 %. This seems to be a realistic aim.

7.7 The Ocean. A Source of Human Food?

7.7.1 How productive is the ocean?

Two thirds of the surface of this planet is covered by the oceans, about 361 million square kilometres.

From Table 7.1 it can be seen that in terms of the biosphere the oceans have the following characteristics:

- The oceans cover 70.7 % of the globe.
- The oceans contain 4 Pg of biomass (dry state) or 0.22 % of the mass of the total global biosphere (see also Figure 7.1).

7.7 The Ocean. A Source of Human Food?

Table 7.2. Productivity of the Ocean

Parameter	Unit	Upwelling and estuaries	Coasts	Open ocean	Total ocean
Surface	Tm^2	0.36	35.0	324	361
Surface, percent	%	0.1	9.9	90	100
Productivity	g/m^2 year	760	250.0	140	155
Productivity	W/m^2	0.40	0.13	0.075	0.076
Annual production	Tg/year	500	8000	46,000	55,300
Number of trophic levels	—	1.5	3	5	4.8
Fish productivity	Tg/year	96	122	1.6	221
	g/m^2 year	270	3.5	0.005	0.61
Fish production in relation to total productivity	%.	32	1.35	0.0034	0.04
The harvest of the ocean					
1950	Tg/year				20
1970	Tg/year				70[a]

[a] *Remarks*:
At the present time the amount of fish caught is 1/3 of global fish productivity and is close to the maximum possible level. The worldwide production of meat (pork, beef, poultry, lamb, etc.) in 1974 was approx. 110 million tons, that is, 110 Tg/year. A large part of this fish harvest is used for the production of fish meal.

- The oceans produce 55.3 Pg of dry biomass per year, that is, 32% of the global production (Figure 7.4).
- This production rate relative to the steady state biomass in the oceans is 13.8 kg/year per kg of static biomass, a factor 215 times higher than that for the continental biosphere.
- The productivity of the oceans per unit of area is 0.153 kg (dry) per m^2 per annum, or 5 times smaller then the continental productivity.
- The oceans' production is not evenly distributed over the whole area. Some areas are highly productive, others very unproductive, as shown in Table 7.2.

It is quite surprising that the open oceans, with 324 million square kilometres, in spite of the large solar energy flux, produce on the average only 0.15 kg/m$^2 \cdot$ year. Continental productivity levels lower than this are found only in the polar regions or in the deserts and semi-deserts (0.09 kg/m$^2 \cdot$ year; Table 7.1).

The average mean productivity of 0.153 kg/m$^2 \cdot$ year means that only 0.08 W/m^2 of solar energy has been transformed. Since the total solar flux is 170 W/m^2, this is an efficiency of 0.05%. 99.95% of the solar flux in the oceans plays no part in photosynthesis.

What is the reason for this? Why are the oceans so unproductive? Table 7.3 gives some of the reasons. It is clear from this that one important factor is the lack of certain essential elements such as iron and phosphorus.

Table 7.3. Constraints on Marine Productivity

Parameter	Unit	Open ocean	Continents	Ratio: ocean continents
Solar energy	W/m^2	170	170	1
Water	Metre of rainfall	"Unlimited" water	0.7	Very big
Mineral contents	kg/m^3	35	> 100	1/3
Surface layer of ocean				
Carbon dioxide in the atmosphere	ppM	~350 Over surface	350	1
Nitrogen in fixed soluble form	g/m^3	0.01	0.04	~1/4
Phosphorus	g/m^3	0.02–0.03	~5000	~10^{-6}
Iron	g/m^3	~ 0.01	>5000	~10^{-6}
Productivity	g/m^2 year	140	785	1/5
Deep ocean layers 1000 m				
Nitrogen in fixed soluble form	g/m^3	0.24–0.35	0.04	~10(?)
Phosphorus	g/m^3	0.05–0.09	>5000	~10^{-5}

The deeper layers of the ocean, below 1000 m, however, contain a higher concentration of these essential elements, so that in those areas where the cold deeper layers well up to the surface the productivity increases dramatically. In these areas the rate of production is some eight times higher than in the open oceans, that is, up to 1000 g/m^2 · year.

Due to the rather complex pyramid of trophic levels, the productivity of the fish population itself varies from 270 g/m^2 · year in these favourable areas to only 0.005 g/m^2 · year in the open oceans.

Covering about 63.5% of the planet's surface, the open oceans' productivity in fish of 0.005 g/m^2 · year would hardly support a reasonable "harvest", even assuming the fish were obliging enough to concentrate themselves into shoals.

Of course, in the lowest trophic level, the green algae, the productivity in the mineral-rich cold water zones is relatively high at 0.2 kg/m^2 year. Of this amount, three quarters is consumed directly by herbivores. The fish in the third and fourth trophic levels have a correspondingly lower productivity level. The real harvest of fish in these areas is not much more than 0.001 kg/m^2 year. Going back to the basic relationship – to that of the primary solar energy

7.7 The Ocean. A Source of Human Food?

– it is found that only 5 ppM of solar energy finds its way into the nets of man in the form of fish meat.

There is some hope for improvement in the future as krill, small crustaceans (8–60 mm long), have a surprisingly large productivity of approx. 1100 kg/hectare. For the sake of illustration, cattle and sheep have a productivity of ~ 800 kg/hectare. Up to 20 kg of krill in 1 m^3 of Antarctic ocean have been measured.

If man were ready to move down the trophic pyramid and begin harvesting krill and plankton, which at present are only consumed by whales, the productivity of the oceans in terms of human food would be dramatically increased.

7.7.2 The ocean is an important source of proteins

The low productivity of the oceans in terms of energy does not diminish the fact that, in terms of food quality in the form of protein-rich food, the ocean is a major resource.

At the present time the contribution of the oceans to humans' food is as follows:

> in energy terms: 2%,
> in total protein: 8%,
> as animal protein: 25%.

In this respect, the position of the oceans in supplying food is much more positive than the earlier discussion on energy may have indicated.

The question must be asked, can this source of animal proteins be increased? These are the most significant and yet the most limiting components in human nutrition.

The answer is not hopeful. The present world catches of fish and other oceanic animals total about 80 Tg/year (80 million tons/year). The total production of the fish population may be estimated as some 220 Tg/year and thus about one third of this is already taken by man. It is likely, therefore, that the true limit is not much higher.

Some commentators postulate that fish productivity could be increased by artificial enrichment of the upper layers of the ocean by mixing with water from the deeper layers, rich in iron, phosphorus, and nitrogen. Due to the obviously high energy requirement for such an activity, others have proposed that the idea is coupled with plans to make use of low-grade thermal power stations floating in the oceans (see Figure 6.10).

There is no question that the ocean is, and will remain, an important source of animal protein. But to imagine that increased fishing alone can solve the problem of feeding a hungry world is not realistic. The solution will have to be found elsewhere.

7.8 Food Production Needs a Large Energy Input

7.8.1 Solar and technological energy input to agriculture

Humans' food is a product of the solar energy flux, but this is not the whole story. Since the beginning of primitive agriculture, cultivation of plants and breeding of animals have relied not only on solar energy but also on the input of human energy (later in the form of energy from domesticated draft animals). Today these energy forces are more than covered by the addition of man-made energy – technical energy. Food production is now a mutual product of all these energy inputs.

The often long pathway from field to mouth gives plenty of opportunity for energy to be applied – work, irrigation, transport, drying, heating, etc. (Figure 7.18).

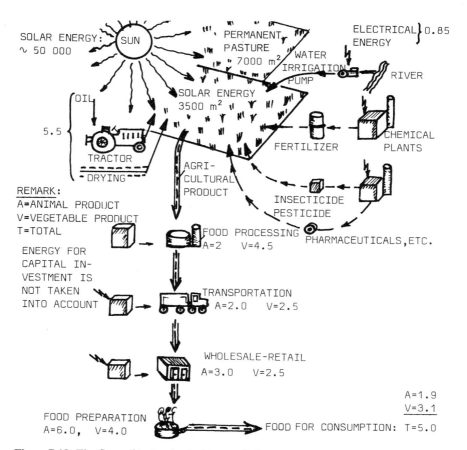

Figure 7.18. The flow of technologically supplied energy needed to produce humans's food is rather complex. (All data in giga-Joule per capita annually.)

7.8 Food Production Needs a Large Energy Input

The inputs are dependent on the level of industrial development in the society. In most developed countries, for direct agricultural application the farmer uses $\sim 10\,\text{GJ}/\text{hectare}$, that is, $1\,\text{MJ}/\text{m}^2$, which corresponds to a continuing energy flux of $0.03\,\text{W}/\text{m}^2$. Since in the developed areas each person requires $3500\,\text{m}^2$ of cultivated land for his needs, the per capita energy input for agriculture alone is 120 W. Much more is used to bring the food via storage processing, preservation, and distribution to the shop. The best estimate for the energy required for these stages is about 600–800 W/capita.

There is a clear tendency now for man-made energy input to agriculture and for food production to increase everywhere. The ratio of technical energy input to produce food "on the table" to the energy value of that food in the highly developed countries is:

$$\frac{\text{energy flux for food production}}{\text{energy flux in food}} = \frac{1000\,\text{W/capita}}{120\,\text{W/capita}} = 8.$$

Figure 7.19 summarises the total energy flux needed for food production.

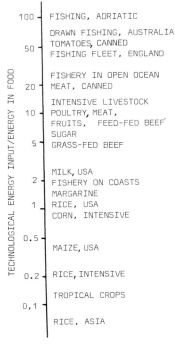

Figure 7.19. Ratio of input of technological energy to the energy in humans' food.

7.8.2 Single-Cell protein – A new food source

Photosynthesis is the primary source of food production. Sola energy is the only source of free energy for biosynthesis. In the near future, however, new methods of food production may be introduced and eventually may play a partial but significant role in the solution to the problem of feeding mankind.

It has been shown that a large amount of man-made energy is needed for food production, particularly production of animal protein, in addition to solar energy. From this it is only a short step to the next idea, which is that the increasing input of man-made energy can result in an intensification of the photosynthesis process and an increase in its efficiency. This can be shown in the case of single-cell organisms which increase under optimum conditions of temperature, input of needed elements such as minerals, carbon dioxide, and water, and the efficient removal of wastes. All this results in a significant increase in production, that is, the output of freshly synthesised dry biomass per unit surface area illuminated by solar light.

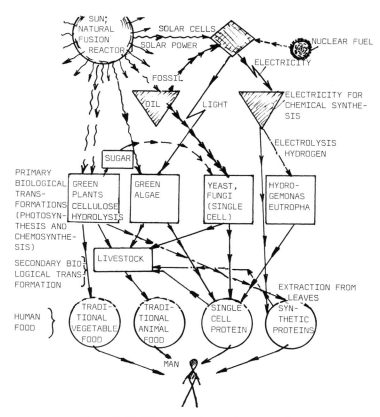

Figure 7.20. New ways to produce proteins.

The next step is self-evident, namely changing the source of free energy. Instead of the solar energy flux one can use the conserved solar energy existing for example in oil reserves. Many microorganisms can "feed" on oil rather than on solar energy (Figure 7.20).

The next step is the replacement of natural oil by synthetic oil based on a man-made, solar independent source of energy – nuclear energy. A further step is the use of man-made light as a source for intensive single-cell photosynthesis. Lastly, the most advanced step is the direct chemical synthesis of biological food for livestock and then for Man by chemistry – the chemosynthesis of amino acids and proteins.

7.9 The Biosphere Is More than a Source of Food

Seen through the eyes of man, the living world is important for the following reasons:

– It is a source of food.
– It is a fundamental element of his own life.

It seems to be clear that the last parameter will play the most significant role in the future. There is no doubt that a population of 8 billion will have to take care of the whole planet, and primarly the biosphere.

It is clear that the number of complex parameters influencing the biosphere is very great. Very deep knowledge of the intimate internal connection between parts of the biosphere, climatic behaviour, and, last but not least, the technological activity of mankind is absolutely essential.

The only question is whether mankind will be ready to fulfill all needs for the protection of the environment.

7.10 What Conclusions Can Be Drawn for Mankind's Future Development?

Taking the points raised in this chapter:

– the number of living beings on the Earth, their mass and the annual production of the biosphere;
– the flow of free energy in the biosphere;
– the production of food for mankind by agriculture;
– the restrictions on further agricultural development and marine farming;
– the coupling of energy flows and the technology of food production;

the following conclusions can be drawn:

- The evolution of the biosphere is closely related to the development of civilisation.
- All the negative influences by man on the biosphere and his environment can be contained by purposeful and properly planned activities in the future.
- The production of sufficient food is possible even with a significantly larger population.
- Continued development of the terrestrial biosphere seems feasible.

These physical restrictions are only one aspect of mankind's development. The real opportunities and choices come from within Man himself.

CHAPTER 8

Is the Future Development of Mankind on This Planet Possible?

> Se non evero, e ben trovato.
> (If it is not true, it is certainly a good lie)
> (Italian proverb)
>
> Being stimulating can be more important than being right.
> M. Rees on F. Hoyle

A student in Cambridge once went to his supervisor in economics and said: "Sir, why do I have to learn all that stuff? I have here the published examination questions for the last ten years, and they are all exactly the same". "Oh yes", was the supervisor's reply, "the questions are the same each year; it is the answers that are different".

T. Gold
(1908–)

The question is often asked, how long would it take a group of how many monkeys working continuously at typewriters to reproduce Shakespeare's "Hamlet"?

You have to realise that some millions of years ago a group of "monkeys" set out on that quest. They not only produced Shakespeare's "Hamlet" but all the other works of Shakespeare, Goethe, Hugo and all the remaining works of literature.

Further, they produced the originals and not reproductions and even invented and built their typewriters!

R. Stratton
(1940–)

8.1 Is It Possible to Consider the Future?

In all previous chapters the past and present states of the material objects in the Universe have been considered – the galaxies, the Solar system, Earth, and human civilisation.

Now it must be asked whether one can look into the future. Is science able to predict future events, future developments? Such questions are too general. They must be very severely limited. The natural sciences have some ability to predict future events but only for well-studied objects and only for simplified cases.

The most complex object known – the collection of human beings called mankind – defies analysis using present-day techniques. There is no basis for objective, scientific considerations concerning the future of mankind. Does this mean there is nothing to be said which is able to shed some light on mankind's future?

There can be no doubt that there exist a number of natural barriers, a number of non-break-through points or ultimate limits. Some of these are listed here (the reader can think of others):

- the limit of velocity, which cannot exceed 3×10^8 m/s;
- evolution along a single direction of time;
- the limit of temperature, which is always greater than $T = 0$ K;
- the conservation of the sum of mass and energy;
- the conservation of baryon number and lepton number (in temperature lower than the Big Bang temperature).

However, these and other physical limits do not directly limit mankind's development.

8.2 The Main Problem: The Increase of the World Population

8.2.1 Is it wrong to consider mankind as part of the biosphere?

The average man has a metabolic energy flux, a power of about 100 W. The present world population of 4.8 billion has a total metabolic power of 4.6×10^{11} W, or 0.4 TW (Figure 8.1).

An exponential increase in the world's population at a rate of 2.2%/year gives a doubling period of 32 years. After approx. 320 years the number will be 1000 times larger, that is to say, mankind will number 4000 billion (or 4×10^{12}) inhabitants. The metabolic flux will equal 4×10^{14} W (or 400 TW). In a further 320 years, that is, 640 years from now, in 2620 A.D., mankind will equal 4×10^{15} inhabitants with a metabolic flux of 4×10^{17} W (or 400,000 TW). The total solar flux absorbed by the Earth today is 122,000 TW.

8.2 The Main Problem: The Increase of the World Population

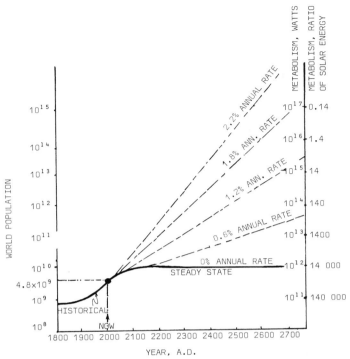

Figure 8.1. World population increase.

Even in the year 2250 A.D., at a 2.2% rate of increase, the metabolic energy flux of the world population would equal the total energy flux in the present biosphere. Decreasing the annual rate from 2.2% to 1.2% has no appreciable effect on the outcome.

A "desired" scenario for the future development of mankind is illustrated by Figure 8.2.

8.2.2 The growth of world population in the past

One hundred thousand years ago Man became master of the world.

Ten thousand years after the Pleistocene Epoch, all regions of the world with the exception of the extreme polar regions were occupied by hunters. At the same time, the sea level all over the world rose as the glaciers of the last ice age, which ended 10,000 years ago, melted, converting many lower coastal and river systems into tidewater estuaries. These coastal estuary systems were extremely rich in marine food resources. Here the first permanent settlements began.

Other settlements arose in the Near East in areas where wild wheat and barley grew naturally and yielded approx. 500 kg/hectare. This allowed a

8 Is the Future Development of Mankind on this Planet Possible?

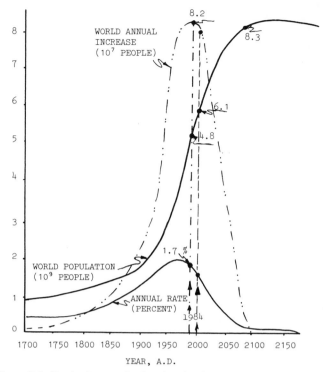

Figure 8.2. Desired scenario for the development of the mankind.

family of four in three weeks of harvesting to collect a year's food supply of wheat and barley.

At this point the increase of the world's population took off. By 8500 B.C. permanent villages existed in the Middle East with well-developed storage facilities. Between 7000 and 3000 B.C., the population in the Middle East increased sixtyfold. This is equivalent to a doubling time of 650 years or an annual increase of 0.11 % (see Figure 8.3).

The next revolutionary step in the increase of the world's population occurred in the 18th century, "the first industrial revolution". The present-day dramatic event is the "second industrial revolution", which puts the annual increase of the world's population on the threshold of 2%.

8.2.3 The reference case used in this chapter – A stable world population of 8 billion

There is no doubt that one of the most important parameters influencing all future events on this planet is the future growth of the world's population. present level of ~4.8 billion will increase in the next 15 years (by the year 2000) to 6.1 billion. This is an annual increase of 1.6%.

8.3 Problem No. 2: A Place on the Earth for Everyone

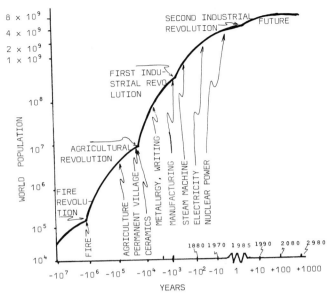

Figure 8.3. The past development of mankind.

As we have seen earlier, this rate of increase leads to a catastrophe in a rather short time. In this book a rather optimistic view is presented whereby mankind over the next 20–30 years will be able to change from a population explosion to a self-limiting development. The aim of this new politics of population growth will be a target of some 8 billion people (Figure 8.2). Of course this target is optimistic, but it does not seem impossible. One cannot, however, exclude other possible scenarios.

It is worth pointing out that even a stable world population of 12 billion instead of 8 billion will not change the prognosis used here in any dramatic way. But the average standard of living will be approx. one-third lower than that for a population of 8 billion.

8.3 Problem No. 2: A Place on the Earth for Everyone

8.3.1 How much space will each inhabitant have in the future?

It is very arrogant to simply divide the total planetary surface by the number of human beings without taking into account the multitude of other living forms, both plants and animals.

About 10,000 years ago, an average of 10,000,000 m² were available to each inhabitant of the Earth. At the present time each inhabitant globally has 38,000 m², equalling an average population density of 27 people per square

230   8 Is the Future Development of Mankind on this Planet Possible?

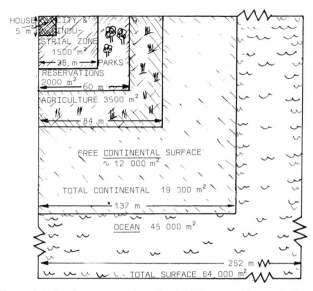

Figure 8.4. Surface per capita (for 8 billion world population).

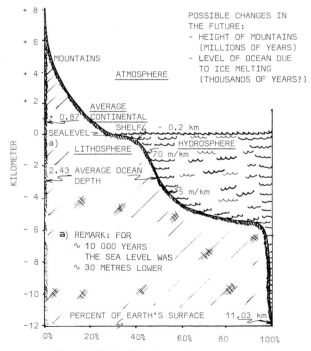

Figure 8.5. Profile of continent and ocean.

kilometre. The majority of the world's population lives on a more restricted continental area with an effective density of 50 persons per square kilometre. In highly developed countries this can rise to 150 per square km. For the postulated world population of 8 billion people the population density will not be dramatically different.

To illustrate the situation in the very distant future, Figure 8.4 gives a very simplified scheme of the distribution of the various regional types over the Earth's surface. A further discussion of this model can be found in section 8.8.

In some projections the expansion of the human population on the bottom of the sea also plays a significant role. There is, however, no serious basis for such a proposal. Even should the tremendous difficulties of living under a pressure of, say, 20 bars (equivalent to a depth of 200 metres) be overcome, the problems lie in other areas.

In Figure 8.5 a profile of the continents is shown. The surface of the continental shelf out to a depth of 200 m is rather small and equals 6.5% of the total surface of the Earth, that is, approx. 33 million square kilometres. Nevertheless, this space could increase the continental space available by 22%. If it could be utilised it would, in the case of an increasing world population, only shift the world catastrophe back by some decades.

The limited surface of the planet is doubtlessly the first and most significant limitation on the world's population.

8.3.2 Organisation of space and transport: The energy lost

There is no doubt that a more-or-less rational organisation of the terrestrial surface has been a significant problem for a very long period of time and will become increasingly so. A random, unorganised use of the surface decreases the order of space, increases the entropy, and results in a sharp decrease of effectiveness. Unfortunately, the rationale for this space organisation is not clear enough for the present society. Historical and irrational influences predominate.

Some general tendencies are obvious. A dramatic increase of the urban population and the growth of a gigantic megalopolis of over 20 million inhabitants is occurring now on every continent. The relationships between the extremes of population density distribution are changing very rapidly.

8.4 Problem No. 3: Food for Everyone

What is the situation regarding food resources for an increasing population?
First some remarks on the principal forms of nutrition in the terrestrial biosphere (see Chapters 4 and 7). Some billion years ago the primary nourishment for the biosphere was the "primordial soup", the aqueous

solution of spontaneously synthesised carbohydrates, amino acids, protovitamins and so on. The simplest organisms depended on these primitive chemical resources available in the primordial soup. Such organisms have been called heterotrophs, that is, those feeding on other "organic" matter.

The big innovation of Nature, the management of chemical synthesis using solar light – photosynthesis – permits the growth of a new class of living organisms drawing their "food" directly from the solar light. These are "autotrophs", the green bacteria, green algae, and plants. This was a major step forward in the evolution of the terrestrial biosphere.

Parallel to the evolution of these autotrophs was that of heterotrophs, which can feed on organic matter, either living or dead, of the autotrophs. Man is a typical heterotroph.

Man himself can cross the boundary of heterotrophy by developing the technological synthesis of food, initially by using biosynthesis, later by pure chemical synthesis (Chapter 7). Before this time arrives, an enormous effort must still be made in the field of traditional food production. Figure 8.6 shows that the possibilities of traditional food production are far from exhausted. A twofold increase of the world's grain production between 1980 and the year 2000 is not impossible.

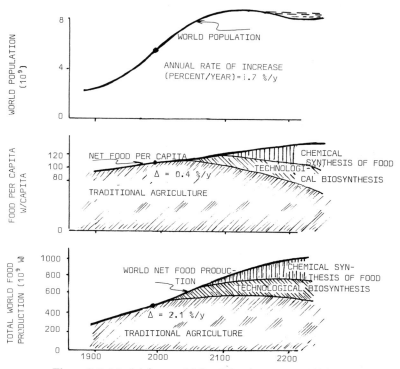

Figure 8.6. Model for world food development – 1984.

A rather rough estimation of the future trends in the world's food production over the next 200 years, during the period of transfer to the steady-state world population, is shown in Figure 8.6. The way to solve the problem of feeding mankind is very difficult, however, and obviously of vital importance.

The changeover from heterotrophs to "technotrophs" (consumption of artificially produced food) cannot be made in the near future, and never completely. A good proportion of Man's food will always be produced by the terrestrial biosphere.

In the meantime the very important problem of the current shortages of food for mankind as a whole can be partially solved by a reduction in the consumption of food from animal sources (particularly meat) and replacement by the consumption of vegetable food rich in amino acids, that is, proteins.

It is not intended here to give details of how this complex problem may be solved. It should be clear, however, that with these various possibilities the problem of supplying the needs of a world population of 8 billion people can be met without dramatic or catastrophic consequences.

It has to be stressed that the production of food at the present time requires inordinate amounts of energy (for machines, transport, distribution, fertilisers, pesticides, etc.). The future energy requirements to supply the necessary food of all types will become increasingly important.

8.5 Problem No. 4: Material Resources for Everyone

8.5.1 Maximum recycling and minimum use

Statements about the Earth's limited resources have been very common in recent years (e.g., "spaceship-Earth"). Many prophets, among them the members of the Club of Rome, have presented complex curves showing the depletion of the planet's mineral reserves. Nearly all of these studies have misunderstood the situation.

We saw in Chapter 5 that the gravitational field of the Earth is the best guarantee for the conservation of almost all of the chemical elements. The few exceptions cannot change this general picture.

It is particularly important to note here that the problem of the carriers of free energy is not being discussed here – the fossil fuels: coal, oil and gas, or even the nuclear fuels, such as thorium, uranium, deuterium, and lithium. These materials belong to the class of energy carriers and will be considered in Section 8.6.

A very rough estimate of the present consumption of materials (including the energy carriers) is shown in Figure 8.7. The total annual material flow is

8 Is the Future Development of Mankind on this Planet Possible?

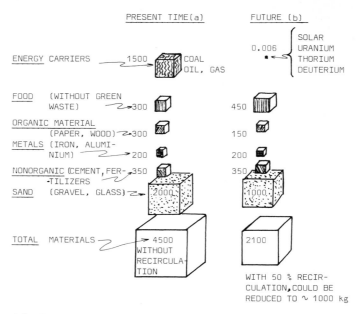

(a) FOR DEVELOPED COUNTRIES
(b) AVERAGE FOR WORLD POPULATION

Figure 8.7. Material flux per capita (kg/capita · year).

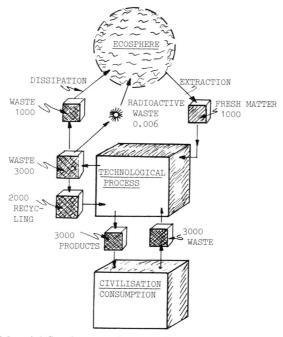

Figure 8.8. Material flux in a steady-state civilisation (in kg/capita, annually).

8.5 Problem No. 4: Material Resources for Everyone

presently approx. 4500 kg/capita. Excluding 1500 kg of energy carriers, approx. 3000 kg/capita is due to the other basic materials.

The amount of materials presently recycled and reused can be taken as zero. Of course, in the future, there are optimistic signs for a more efficient management of the use of materials which will have the effect of decreasing the annual flows to a level of approx. 1000 kg (including the energy carriers).

It is extremely arrogant to try to give a picture of the future complex distribution of material flows. Nevertheless, an attempt is made in Figure 8.8,

Table 8.1. Possible Annual Material Usage per Capita in the Future

Element	Content in one ton of ecosphere		Potential production, kg			Nuclear fuel	Other products and wastes
	mol	kg	"Metals"	Cement, glass	Ferti-lisers		
Oxygen	31,600	500		Z(80)			300 free oxygen
Silicon	9180	257	60 (Si_4N_3)	300 (SiO_2)	—	—	(SiC_xH_y)
Hydrogen	720	0.13	—	—	—	—	—
Deuterium	0.13	0.0025	—	—	—	—	—
Aluminium	2900	78	78	—	—	—	—
Sodium	1050	24	—	Z(24)	—	—	—
Potassium	780	33	—	Z(30)	3	—	—
Iron	840	47	47	—	—	—	—
Magnesium	700	17	17	Z(10)	—	—	—
Calcium	575	23	—	Z(18)	5	—	—
Titanium	85	3.8	3.8	—	—	—	—
Phosphorus	34	1.1	—	—	1.1	—	—
Fluorine	31	0.6	—	—	—	—	—
Carbon	16	0.2	—	—	—	—	—
Nitrogen	10	0.15	—	—	0.5+air	—	—
Sulphur	8	0.25	—	—	0.25	—	—
Uranium Thorium Lithium	—	0.1				0.1	—
Other	37	1.5	—	—	—	—	Wastes 45
Total	55,650 mol	1000 kg	Metals 135 kg Si-nitride 60 kg	(Z=cement) Cement 150 kg Quartz Glass ~300 kg	Ferti-lizer 10 kg	0.1 kg, which corresponds ~10^{13} J	345 kg

but the absolute values shown must be regarded as completely speculative. It shows, however, that the only basis for a discussion of the future of mankind includes a large degree of fantasy!

The postulated material flow is based on the average chemical composition of the Earth's crust and the potential possibilities for using different elements for different technological purposes. Such roughly estimated material flow is given in Table 8.1. Of course, in all these considerations the use of water and air has been neglected.

8.5.2 Material recycling and energy

The estimates given in Figure 8.9 permit a forecast for the annual material flows for mankind for a stable civilisation of approx. 8 billion.

The annual material flows, after taking into account effective recycling, can be reduced to 16×10^{12} kg/year. This equals a volume of approx. 8 cubic kilometres, that is, a cube with a side of 2.0 km. It seems that such an amount of material taken from Earth's crust will be possible even in the distant future.

After one million years of such a civilisation with this material flux, the surface of the Earth will be reduced by an equivalent depth of 2.0 km over an area of 4 million square kilometres, that is, a square having a side of 2000 km. Such an exploited, but not completely destroyed, surface will of course be

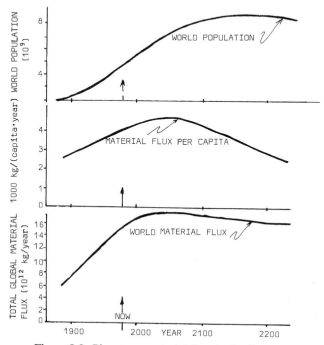

Figure 8.9. Planetary material flux in the future.

distributed over the surface of the globe. Only approx. 4% of the continents, 1.2% of the whole planet, will be so affected – one million years from now. In all of this, consideration of the deeper exploration of the Earth's crust, say to depths of 20–25 km or even approaching the Earth's mantle, has been neglected.

8.6 The Ultimate Problem for the Future of Mankind: The Flow of Free Energy

8.6.1 Why energy?

The most important problems for future civilisations have been mentioned earlier:

- organisation of the Earth's surface;
- population distribution;
- transport organisation;
- food production including the technological synthesis of food;
- material resources and the large problem of recycling materials;
- environmental protection and climate regulation.

There are, of course, other problems:

- health and hygiene,
- scientific, cultural, and spiritual development, and
- organisation of reserves.

In this book we are limited to problems of matter and energy. These can be resolved into two principal problems:

- the limitation of the world population as is postulated at a maximum of 8 billion, and
- securing the consistent continuous flow of free energy.

If a satisfactory supply of free energy can be ensured, then all the other problems can also be solved.

8.6.2 The prognosis for energy consumption

Chapter 6 outlined the amount of energy needed by the average human being for a good standard of living. More or less arbitrarily this was set at a level of 8 kW of primary energy per capita. For the sake of illustration some present power levels per capita are as follows (remark: 1 kW corresponds to 1 kW year/year, that is, 31.5 GJ/year):

India   1   kW/capita
World average   2.3 kW/capita
Federal Republic of Germany   7   kW/capita
USA   12   kW/capita.

Further, it will be assumed here that the world average will reach 8 kW in approx. 150–200 years and that at that time the world's population will have reached a steady state.

The question arises, how much energy will Man need in the future? Figure 8.10 gives a very simple but self-explanatory answer. The relative and absolute changes are shown.

Now enough information or assumptions are available for preparing the prognosis for the future needs of a steady-state civilisation. The assumptions are as follows:

– The world population levels off at 8 billion.
– The average energy flux increases to 8 kW/capita.
– The total energy needs for technology rise from a present level of 11 TW (4.8 billion people with an average 2.3 kW/capita) up to $\sim$ 60 TW (8 billion with 8 kW/capita).

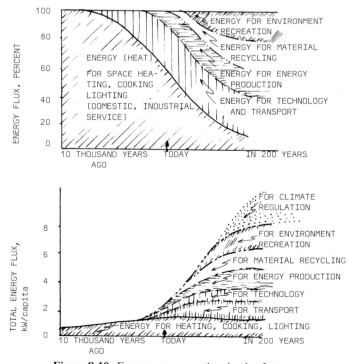

Figure 8.10. Energy consumption in the future.

8.6 The Ultimate Problem for the Future of Mankind: The Flow of Free Energy

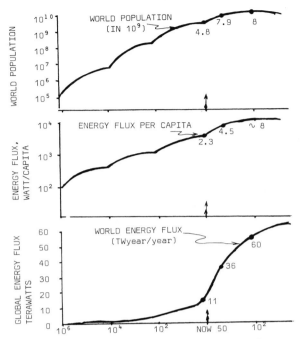

Figure 8.11. Technological energy flux in the future.

There is no doubt that the energy consumption rate here is overestimated, but it seems to be feasible. This data is given in Figure 8.11.

It is clear that in all this discussion the term "energy" or "technological energy" means the flow of energy which is produced, transformed, stored, and used in man-made devices in technological processes. The huge amount of the natural flux of solar energy reaching, for example, agriculture and forests has not been taken into account. Here we must be careful, since in the course of time the share of solar energy used in technological ways must increase.

It also has to be stressed that the energy needed for future climatic regulation has been neglected in spite of the possibility that it could be a significant amount.

8.6.3 How much energy is needed to produce the technological energy used by Man?

There are two different fluxes of energy in civilisation:

— the energy flux for maintaining order in society, food for Man, building and heating of houses, transport of materials and people, production of useful things, maintenance of science and culture, and maintenance of social and political order;

- energy flux used to produce energy, for example, for recovering crude oil, for pipelines and tankers, for construction and maintenance of oil refineries and power plants, for transport (transmission) and storage of energy carriers, and for management of the resulting wastes.

The first part of the energy flux is really needed for the development and maintenance of Man's well-being; the second is only the energy "price" to be paid for meeting these energy needs. If it were possible we would do without this second part – but it is not.

"Free" energy is found in the stored energy contained in coal, oil, hot rocks, uranium ore, deuterium in water, etc., or in the form of energy flux (solar light, wind flux or geothermal heat). All of these are provided "free of charge", but the process of transforming this freely provided energy into a useful form for the consumer is complicated, needing energy and equipment to realise it.

Consider the simplest case, the use of wood in prehistoric times. Wood was spatially available close to prehistoric man. Even here some human energy, some human work, was required for collecting, transporting, chopping, and storage. Even this "free fuel" was not free of charge, when the cost is expressed in energy units.

8.6.4 The future source of free energy

The number of different and often even contradictory forecasts concerning the future sources of free energy is rather large. An attempt has been made here to split the alternatives between the known and more-or-less established energy sources, including fusion technology.

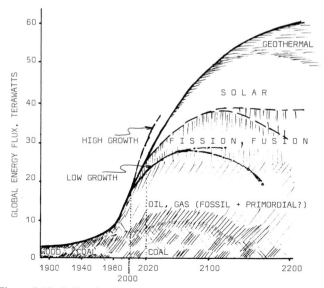

Figure 8.12. Split of energy production over the next 200 years.

8.6 The Ultimate Problem for the Future of Mankind: The Flow of Free Energy

Table 8.2. The Energy Sources in the Far Future, Arbitrary Split (in Steady-State Civilisation)

Type of energy	Primary energy flux		Energy per year	The energy carrier or absorber	
	TW Annual average	%	Terajoule year	Energy content (J/kg)	Total amount (kg/year)
Total	60	100	1.9×10^9	—	—
Fossil fuel (oil, gas, coal)	3	5.0	9.5×10^7	2.9×10^7	3.2×10^{12}
Fission (Uranium, Thorium)	20	33.3	6.3×10^8	8.6×10^{13}	7.4×10^6
Fusion (Deuterium, Lithium)	15	25	4.8×10^8	3.5×10^{14}	1.4×10^6
Geothermal [a]	1	1.7	3.2×10^7	6.3×10^5	5×10^{13}
Hydropower [b]	2	3.3	6.3×10^6	8×10^5	8×10^{12}
Solar energy [c]	19	31.7	6.0×10^8	$\sim 600{,}000$ km² solar absorber	

[a] Temperature range: 180 °C → 30 °C.
[b] Height 100 m: efficiency 0.8.
[c] Average annual power 30 W (el)/m² of absorber.

Figure 8.12 gives a possible split of energy production over the next 200 years. Of course, the value of all such forecasts is questionable. The only positive statement which can be made is that the existing technologies are capable of further development and allow a qualitative increase in mankind's well-being.

A rather rough estimate of the fuels required – coal, oil, uranium, thorium, deuterium, and remarkable resources such as solar, geothermal, and hydropower – are shown in Table 8.2. Here again the data are very uncertain.

8.6.5 Not only the free energy sources are important but also the sinks!

In the arbitrary scenario chosen here, the future flow of technological energy has been set at 60 TW. Of this, ~ 20 TW comes from solar energy, leaving 40 TW from other sources.

The total energy flux of the planet at the present time (as was mentioned in Chapter 6) is 122,000 TW. The additional flow of technological energy of non-solar origin will only be a relatively small part of the total:

$$\frac{40 \text{ TW}}{122{,}000 \text{ TW}} = 0.00033 = 0.033\,\%.$$

Such a small change in the energy balance of the Earth is doubtlessly insignificant. The changes in the primary energy flux from the Sun are much more significant (see Chapter 6).

The remaining very important problem results from the local impact of waste heat being produced by a large power complex. Even a sophisticated power station of 2 GW (el) will also, in the future, have a heat waste of approx. 2 GW (thermal). If it is assumed that this waste is released over a surface of 100×100 m, then the local power density will be 2×10^5 W/m^2, or over 1000 times higher than the intensity of the average solar flux and 200 times higher than the solar energy flux at midday.

8.7 The Future Climate of This Planet

8.7.1 Will the terrestrial climate remain favourable?

The development of mankind stretching into the far future has been discussed, without taking into account the effect of climatic changes. Only the so-called "greenhouse effect" due to man's production of carbon dioxide was touched on.

An even more fundamental question can be formulated: Will Man be able to control the planetary climate so well that the coming Ice Age could be reduced in its intensity or even prevented?

The "climasphere" is an enormous thermodynamic machine with an energy flow of approx. 122,000 TW. But to control this machine, to steer it, perhaps only one thousandth or one ten-thousandth of this energy flux is required, that is, only 122 TW or even 12 TW. The answer lies in the further development of climatology and of the physics of the atmo-hydro-lithosphere. Should it prove possible to control the climate, then Man will have a major tool in his hand for his further development. The highest importance must be given to

- controlling the climate, and
- accounting for the additional 10–100 TW which might be needed to do this (Figure 8.13).

In all the considerations of energy requirements, the need for climate control has been neglected. The energy flux for the "classical" human activity has been assumed to be 60 TW. Additionally, we must add 10–100 TW for climate control, that is, the same amount again, giving a possible future requirement of 120–150 TW.

It is very important to note that the recent modelling of climatic evolution predicts that the long-term cooling trend which began some 6000 years ago will continue for the next $\sim 20,000$ years.

8.8 The Quality of Life 243

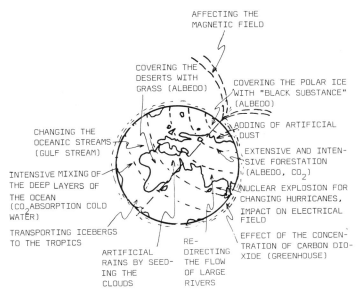

Figure 8.13. The possible ways of influencing the planetary climate.

8.7.2 The possibility of controlling the terrestrial climate

The idea that mankind will be faced with the problem of controlling the climate in the distant future has consequences for the increase in energy requirements. Of course, such a situation can only occur sometime after the next 200 years. Such a scenario is given in Figure 8.13.

The problem arising from this is that an additional source of energy of some tens of terawatts has to be found. The simplest answer is to assume that nuclear energy, either fission or fusion, will supply this need. From Table 8.2 it can be seen that even doubling the total demand for 60–120 TW, allowing 60 TW for climate control, will mean the annual consumption of 22,000 tons of uranium and thorium (with the breeder) or 5600 tons of deuterium and lithium, or a combination of both.

8.8 The Quality of Life

All the parameters discussed above are only a small part of all those factors which may influence Man's future, his well-being, his standard of living, and his happiness.

It is not the aim of this book to discuss all these factors, but some of the essential assumptions concerning a model of a region with a population of 1 million inhabitants have been discussed here. The more or less arbitrarily chosen assumptions are the following:

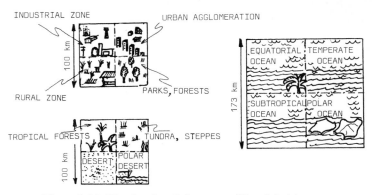

Figure 8.14. The "regions" for one million inhabitants.

- highly industrialised and urbanised areas of 5000 km² (Figure 8.14), including a central power station complex of 8 GW (total), which corresponds to 8 kW/capita;
- rural area of 5000 km², which corresponds to 5000 m² of agricultural area per capita (see Figure 7.19);
- distant vacation areas – rest and recuperation – including tundra, desert, and polar regions;
- free oceans with an area of 30,000 km² and with 100 km² of artificial islands;
- rational food production by agricultural, biosynthetic, and chemosynthetic methods;
- the recycling of materials;
- the use of solar and nuclear energy at a level of 60–200 TW, depending on whether climatic controls are needed.

8.9 What Conclusions Can Be Drawn Concerning the Future Development of Mankind?

From all the questions discussed in this chapter:

- the necessity to limit the human population at around 8 billion;
- the consequent achievement of a stable population level;
- better organisation of the use of the Earth's surface for human activity, that is, agriculture and preservation of natural resources;

it can be seen that:

- There are no fundamental reasons why a stable population of 8 billion should not exist on this planet for a very long time.

8.9 What Conclusions Can Be Drawn Concerning the Future Development of Mankind? 245

- This stable state could be readily reached over the next 200 years.
- The transition period must, however, begin now and considerable effort and organisation must be applied to achieve it.

As a final comment, it is noted that this is only the physical basis for mankind's future development – the ultimate decision and the methods chosen lie with Man himself.

CHAPTER 9

The Distant Future of Mankind – Terrestrial or Cosmic?

What really interests me is whether God had any choice in the creation of the world.

A. Einstein
(1879–1955)

It is very difficult to make an accurate prediction, especially about the future.

Niels Bohr
(1885–1962)

Our society is built not on the joy and happiness of the past, but on the agonies experienced by the long line of our predecessors. Whether or not all the agonies and struggles of the past will emerge into a great future, or will vanish into nothing at all, is likely to be decided in the next few tens of human generations.

Fred Hoyle
(1915–)

What can a stone-age man expect from extraterrestrial activity? Probably an enormous, excellently polished stone tool! And the antique man? A big galley with a kilometre-long oar. And the modern man? A civilization with an extremely high flux of energy!

Stanislaw Lem
(1929–)

This Universe is our only chance, and we had best make the most of it.

J. A. Wheeler
(1911–)

If there's intelligent life out there, why haven't they visited us? They're too intelligent!

Sky and Telescope, 1981

9.1 The Natural Constants and the Future of the Universe

9.1.1 The very far future

The development of civilisation over the next 200 years has been discussed in the previous chapter. From this it could be seen that it was not unreasonable to hope for a rather smooth and continuing development over that period. No sudden changes are foreseen when one thinks about planetary and cosmic parameters. More dramatic changes could, however, occur in human conditions in the next hundred years. The present chapter will look at a much larger time scale – the very distant future – thousands, millions, or even billions of years away. Does one have the ability or a reason to look so far into the future? What can be gained from such an excercise?

9.1.2 How stable are the natural laws and constants?

The most fundamental question concerns the basic structure of the natural world in the distant future:

- the stability of the natural laws: law of conservation of energy and mass, conservation of baryon and lepton numbers;
- the stability of the relative strengths of the elementary interactions, which at present are:

  - strong interaction ~ 1
  - electromagnetic interaction $1/137 \sim 10^{-2}$
  - weak interaction 10^{-12}
  - gravitational interaction 10^{-39}

- the stability of the unit of electrical charge and the quantitative equality of the negative electrical charge of the electron and the positive charge of the proton;
- the stability of the natural constants, for example,
  - the velocity of light $c = 2.9979 \times 10^8$ m/s
  - the gravitational constant $G = 6.673 \times 10^{-11}$ Nm^2/kg^2

  (see Table 9.1).

These and similar questions can be posed. There are indeed some hypotheses concerning the dependence of these laws and natural constants on time. This means that not only in the past, but also in the future these values may well be different. In this chapter it is assumed, nevertheless, that all remain unchanged for the very distant future.

Table 9.1. How Constant Are the Natural Constants and Laws?

Problem	Content	Limit	Reference
Is the proton stable? ($t_{1/2}$=half-life)	Baryon number stability $p^+ \to \pi + e^+$	$t_{1/2} > 10^{32}$ years	Experiments, 1983
Is the electron stable?	Lepton number stability $e^- \xrightarrow{???}$	$t_{1/2} > 10^{21}$ years, probably electron is stable	
Is the electrical charge symmetrical? ($\Delta\varepsilon = E_{p+} - E_{e-}$)	$|E_{p-}| = |E_{e-}|$	$\Delta\varepsilon < 10^{-21}$ the electrical charge of the proton is exactly equal to that of the electron	Hughes, 1963
Is the coulomb law quadratic? (r=distance)	$F \propto \dfrac{1}{r^{(2+q)}}$	$q = (2.7 \pm 3.1) \times 10^{-16}$ quadratic law is consequence of three dimensional space	Williams, 1971
Is the elementary charge, e, constant? (C = coulomb)	$e = 1.602 \times 10^{-19}$ C	$\dfrac{\Delta e}{e} \leq \dfrac{1}{1600}$ in 10^9 year	Dyson, 1960
Is the fine structure constant, α, constant? (e=electrical charge, $\hbar = h/2\pi$)	$\alpha = \dfrac{e^2}{\hbar \cdot c} = \dfrac{1}{137.03595}$	constant for distant galaxies with red shift $z = 0.2$	Bahcall, 1967 Tubs, Wolfe, 1980
Is the gravitational constant, G, constant?	$G = 6.673 \times 10^{-11} \dfrac{\text{Nm}^2}{\text{kg}^2}$	$\dfrac{1}{G}\left(\dfrac{\Delta G}{\Delta t}\right) < 10^{-11}$/year	van Flanderin, 1981
Is the mass of the electron stable?	$m_e = 9.10954 \times 10^{-31}$ kg	?	
Is the mass relationship stable? (m_p=mass of proton; m_e=mass of electron)	$\left|\dfrac{m_p}{m_e}\right| = 1836.151 \pm 0.43 \times 10^{-6}$	?	Tubs, Wolfe, 1980
Is Planck's constant, h, constant?	$h = 6.626189 \times 10^{-34}$ Js	$\dfrac{1}{h}\cdot\left(\dfrac{\Delta h}{\Delta t}\right) < 10^{-18}$/year	Pegg, 1977
Is the velocity of light, c, constant?	$c = 2.998 \times 10^8$ m/s	?	?

9.2 The Future Development of the Universe

The history of the Universe from the Big Bang to the present moment, that is, approx. 12 billion years, has been full of dramatic events (see Chapter 2).

This was in the past. And the future? Can past history be extrapolated into the future? How can the course of history be changed by the variation of some natural constants? (See Table 9.2.) It is enough to know the most important parameters of the present Universe, the density of matter and energy, in order to form a reasonable idea about the Universe's future evolution.

Table 9.2. Universal Forces and Properties and the Results of Varying Their Values

Phenomenon, Value	Consequences of:	
	Lowering	Raising
Strong $\sim 1/3$	No nuclear reactions at all. No heavy elements made.	Early Universe converts all matter to heavy elements; no source of energy for stars.
Electromagnetic $\sim 10^{-2}$	Electron not bound to atoms no chemical reactions possible.	Electron inside nucleus: No chemical reactions possible.
Weak $\sim 10^{-12}$	No hydrogen "burning" possible: No source for energy to stars: $4H \rightarrow {}^4He + 2e^+ + 2\nu$	Early Universe converts all matter to helium. No source of energy for stars.
Gravitation $\sim 10^{-39}$	Stars do not get hot enough for nuclear reactions to occur.	Nuclear reactions so rapid, that star lifetime is very short.
Expansion (Hubble's constant) $H_0 = 17 \frac{km/s}{\text{Mega light-year}}$	Matter comes out of early Universe in dense configurations.	Galaxies cannot form, matter ends up spread out uniformly.
Density of Universe $\varrho_0 \simeq 10^{-27} \frac{kg}{m^3}$	Galaxies cannot form.	Early Universe turns all matter into heavy elements.
Temperature $T_0 = 2.9 K$	Early Universe turns all matter into heavy elements.	Galaxies cannot form due to radiation pressure.
Structure: isotropy, homogeneity.	Galaxies cannot form in anisotropic expansion.	

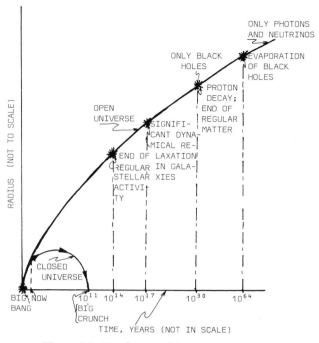

Figure 9.1. The future of the open universe.

If the density of matter and energy is less than 8×10^{-27} kg/m³, the Universe will expand to infinity. For a Universe with only a slightly higher density, for example, with twice that density, the future will be significantly different. After some tens of billions of years of further expansion it will slow down and stop and then begin to contract and, after further tens of billions of years, again reach a fully collapsed state – the Big Crunch! (See Figure 9.1).

But even the "open" Universe, which expands without limit, includes the death of stars, of matter (because of proton instability), and even the end of the black holes ("evaporation" process) (Figure 9.1).

In each case the Universe will be stable for at least another ~ 100 billion years or so – there is no cause for concern!

9.3 The Future of the Galaxy and the Sun

9.3.1 The stability of galaxies

Are galaxies as stable as the Universe? Surely not. The galaxies are younger than the Universe, possibly some millions of years younger. There are at least 100 billion of them. There is a wide range of types, differing in age and evolutionary paths. Some theories assume that there may also be galaxies of

9.3 The Future of the Galaxy and the Sun

Table 9.3. Catastrophic Events in Our Region

Event and total energy emission (PJ = 10^{15} J)	Frequency (years)	Energy arriving in the upper atmosphere (PJ)	Kind of energy	Possible direct irradiation of man during his life	Secondary impact on terrestrial parameter
Explosion in galactic centre: 10^{31}–10^{36} PJ	Once per 500 million years	$\sim 10^6$ PJ	Gamma rays		In most papers concerning these possible effects, the destruction of ozone layers (shielding against the solar ultraviolet light) is mentioned as the most important event, especially when it is combined with the changes in terrestrial magnetic fields.
Supernova total energy: 10^{33} PJ	Once per 30 years in the Milky Way; once per hundred million years within ~ 60 light years	10^6 PJ	Cosmic rays	1000 Joules per human body during 80 years	
Supernova expanding shells: 10^{30} PJ	Once per several hundred million years	$< 10^4$ PJ	Kinetic energy of gas and dust cloud		
Large solar flare: 10^{13} PJ	Once per several thousand years (?)	10^6 PJ	Cosmic rays	1000 Joules per human body during short time.	
Comet, meteorite $\sim 10^5$ PJ	Once per 28 million years	10^5 PJ	Body fall: kinetic energy	—	See also Figure 7.9
Apollo objects (~ 1 km diam.)	Once per 250,000 years	4×10^5 PJ		—	Crater of ≈ 20 km
Cretaceous comet	10^{15} kg	$\sim 10^9$ PJ	Chemical poison?	—	See also Figure 7.9

Remark: 100 rad = 1 Gr = 1 J/kg
Human body: ≈ 70 kg
Lethal dose: ≈ 500 rad = 5 Gr/kg
Lethal dose for human body: 5 Gr/kg × 70 kg = 350 Gr = 350 J
Here for simplification: Lethal dose for human body during 80 years $\cong 1000$ J

dead stars, of cold black dwarfs, and even of black holes. If this were confirmed the consequence would be one model for short-lived and another for long-lived galaxies. The galaxy in which the Earth is situated, the Milky Way, has a rather typical spiral shape and is made up of both old and young stars. The Milky Way is doubtlessly far from being old and is certainly not dying.

Is the Galaxy old enough to have reached an inactive stage? In reality the opposite seems to be true. The Galactic centre is rather active and this is manifested by very dramatic events. At least part of the astrophysical community has the opinion that the energy of an explosion in the Galactic centre must be of the order of 10^{51} J, the conversion of a mass of more than 10,000 Suns completely into energy. The mass of material ejected must be more than 100 million times the mass of the Sun.

Such explosions may occur once every 500 million years. The total energy required for explosions over the lifetime of a galaxy comes to 10^{53} joules and the total mass ejected from the centre is 10^9 solar masses (Table 9.3).

Another explanation reduces the explosion energy to 10^{45} J, and then the ejected mass would have to be 100,000 solar masses. Here again the impact on the stars near the Galactic centre can be very significant.

The Solar system is quite far from the Galactic centre, approx. 30,000 light-years. Is the Sun out of danger?

9.3.2 How stable is the cosmic neighbourhood of the Solar system?

The centre of the Galaxy is sufficiently far away, but there are dangerous objects in the galactic neighbourhood. Of course, they cannot release as much energy as an exploding Galactic centre, but, on the other hand, they are much nearer. They are the supernovae.

A typical supernova (see Chapter 3) emits, over a relatively short period of several years, a total of approx. 10^{45} J. Assuming that the distance to the Solar system is, say, 35 light-years, a simple calculation can be made:

Energy reaching the Earth's upper atmosphere

$$\cong \frac{10^{45} \text{ J}}{\left(35 \text{ years} \times \left(3.15 \times 10^7 \frac{\text{s}}{\text{year}}\right) \times \left(3 \times 10^8 \frac{\text{m}}{\text{s}}\right)\right)^2} \cong 10^9 \text{ J/m}^2.$$

It must be stressed that the energy flux reaching the Earth is not experienced after 35 years but is spread over a much longer period of ~ 80 years due to the screw-like movement of emitted particles along the interstellar magnetic field lines. Some rough estimates concerning typical cosmic catastrophes and their impact on the Earth are given in Table 9.3. The probability of these cosmic catastrophes is summarised in Figure 9.2.

9.3 The Future of the Galaxy and the Sun

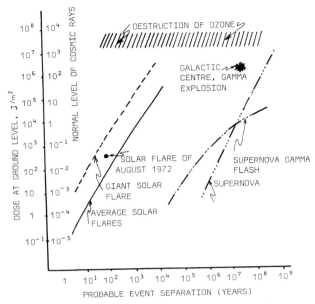

Figure 9.2. Probability of cosmic catastrophes.

In all these considerations it must be noted that, in the last 65 million years, the Earth has not experienced any effects of a "cosmic" catastrophe and in particular, no event has been strong enough to cause major changes in the biosphere or climasphere. This does not mean, however, that the problem of cosmic catastrophes can be neglected.

The Sun rotates around the Galactic centre, and in the life of the Earth (4.5×10^9 y) the Sun has made approx. 20 galactic revolutions. No supernova has destroyed the Solar planetary system. It can therefore be assumed that, in a further 20–100 galactic revolutions, a supernova explosion will not occur within 100 light-years.

It is worth noting that modern astronomy has shown that, as far as gamma rays are concerned, Man's planet lies in a relatively quiet galactic suburb. Most of the activity – new stars, supernova explosions, pulsars, and cosmic rays – are found thousands of light-years away, much closer to the Galactic centre. The Earth is at one of the quiet outer edges of that vast system known as the Milky Way, therefore our house – the Solar house – is a surprisingly safe home!

For more than a century mankind was convinced that the Solar system was rather stable and safe and that at least from this direction no danger threatened. In recent years this belief has been changed dramatically. The following mechanisms threaten the future of our planet.

The Earth is most likely influenced by the fall of large comets and/or meteorites and passes through gas-dust clouds which are quite numerous on the galactic plane. These "catastrophes" seem to be reccurrent, with a period of approx. 28 ± 2 million years. There are two independent causes of these catastrophes:

1) A small companion of the Sun, a brown dwarf having a mass of 0.01 solar mass, on a highly eccentric elliptical orbit, which encounters the "comet ring" once every 28 million years and might be responsible for sending comets to the Earth.
2) The oscillation of the Sun perpendicular to the Galactic plane, resulting in the crossing of the plane once every 28 million years, may cause gas and dust to come between the Sun and the Earth, resulting in climatic change. Also, the increase of the fall of comets could be influenced by this mechanism.

Most of these hypotheses agree that a recent catastrophe destroyed a large number of species approx. 13 million years ago. The extinction of dinosaurs and other species 65 million years ago may have had the same cause.

According to these hypotheses, in 15 million years mankind can expect a new catastrophe to the terrestrial biosphere, and therefore a catastrophe for mankind himself.

It seems that for future civilisations such a catastrophe will be not unmanageable.

9.3.3 How stable, how predictable is the Sun?

The central star of the Solar System, the Sun, is a typical star of the third generation, a typical dwarf of the Main Sequence (see Chapter 3).

The Sun, being 5 billion years old, is probably in the middle of its life. A further 4–5 billion years will pass before the Sun leaves the Main Sequence and becomes a Red Giant (Figure 9.3). For the next few billion years, therefore, the luminosity – that is, the energy flux of 4×10^{26} W, and the surface temperature of 6000 K, corresponding to the emission of visible light – will remain the same as today.

One of the fatal constraints to the long-term existence of mankind on the Earth is the threat of the transformation of the Sun from a main sequence star to a red giant star and finally contraction to a white dwarf star (see Figure 9.3).

One possibility for survival might be to shift the Earth by rocket propulsion from its present orbit to that of Jupiter. Such a shift needs a power of $\sim 10^{19}$ W, which could be produced by burning deuterium at a rate of 2.4 ton/sec and by using hydrogen as a propellant at a rate of 15,000 ton/sec. The proportion of deuterium/hydrogen required, 1:6500, corresponds roughly to the isotopic composition prevailing on Jupiter.

The fuel and propellant would have to be transported from Jupiter by rocket techniques. The time when the orbit transfer would need to take place is in approx. 4 gigayears and would take approx. 1 gigayear to accomplish.

9.3 The Future of the Galaxy and the Sun

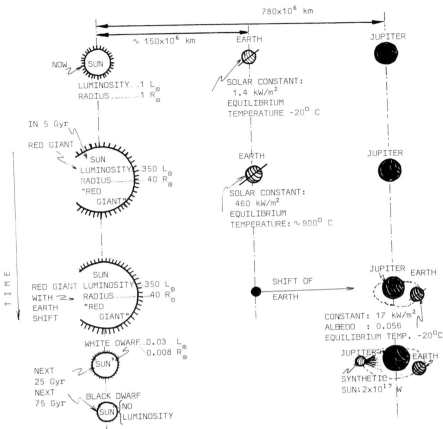

Figure 9.3. Evolution of the Sun and shifting of the Earth.

Jupiter could also serve as a source of deuterium for a "synthetic" Sun, which could provide an effective flux of $\sim 1.7 \times 10^{18}$ W on the Earth's surface lasting 100 gigayears. This would exhaust about 8% of Jupiter's deuterium reserves (see Chapter 4). After this period, if one accepts the hypothesis of the closed Universe, the "Big Crunch" finishes the story.

The Sun, like other stars in the Main Sequence, shows some instabilities. The best-investigated phenomena are the solar eruptions, the solar flares. During normal or quiet periods the solar flares show rather small energies, on the order of magnitude of 10^{10} PJ for a typical period of several hours. There are reasons to believe, however, that once in a thousand or ten thousand years the solar flare's magnitude can be 100 or even 1000 times greater. In this case the direct impact on the Earth and its biosphere would be far from insignificant (Table 9.3; Figure 9.2).

9.4 The Future of the Planet Earth

9.4.1 The stability of the planet

Having seen that the Sun is a rather stable star, one must next ask: Is the planetary system as stable as the central star? How stable is the planetary system? How stable is the Earth itself?

The existing models of the planetary system provide strong grounds for saying that the planetary system is very stable, with a long life expectancy and with a similar stability in the relative position of its principal members.

Given that the Earth itself is relatively stable as a cosmic object, how stable are the constituents of the planet? (Figure 9.4):

- the lithosphere, affected by erosion, continental drift, a source of heat coming from radioactive decay;
- the atmosphere, a product of the activity of the biosphere, oxygen from photosynthesis, nitrogen also perhaps a product of the biosphere;

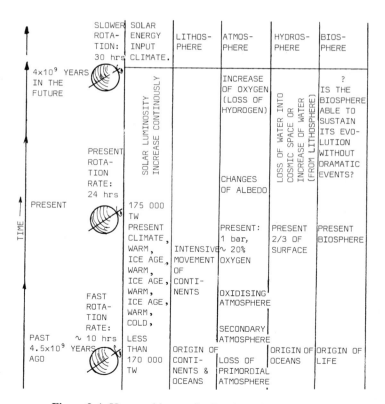

Figure 9.4. How stable are the Earth and its climate?

9.4 The Future of the Planet Earth

- the hydrosphere, being destroyed by photolysis due to ultraviolet radiation in the upper layers of the atmosphere and then depleted by escaping hydrogen;
- the biosphere as a planetary parameter being an object of continuous evolution.

The only objective answer to this question is that, for at least the next few billion years, the Earth and its constituents will remain fairly unchanged.

9.4.2 The fall of small cosmic objects and earthquakes

There are other dangerous objects in cosmic space or more particularly in our Solar System. Numerous meteorites of various sizes and comets (remnants of protostar system?) move in the vicinity of the Earth, and direct encounter and impact with the Earth's surface occurs occasionally.

On June 30, 1908, near the Siberian village of Tunguska, a comet (?) with a mass of 3×10^7 kg struck the Earth. A gigantic "impacted area" can be seen even today. The energy released during this impact has been estimated at 50 PJ. Many people observed this cosmic event and many scientific measurements have been made. Even this event, however, had only a rather local and limited effect.

The probability of such an impact on the Earth is about one every thousand years. In the past, many similar events have occurred but no significant damage has been left behind apart from a few small craters.

There are some theories that the Cretaceous extinction of the dinosaurs resulted from the fall of an asteroid with a mass of 10^{14} kg (~ 7 km diameter), which corresponds to an energy release equivalent to the explosion of 100 million megaton bombs.

The Earth itself is also a source of tremendous energy releases in the form of earthquakes and volcanic eruptions. The best investigated volcanic explosion is that of Krakatoa, which occurred on August 26, 1883, in the Sunda Strait. This catastrophic event released energy equalling 1000 PJ and ejected material totalling 18 km³ or 50×10^{12} kg. Some of this material, in the form of ash, was propelled to a height of 80 km, even higher than the ozone layers. In 1500 B.C., in the Aegean Sea, the volcano Santorini released an energy of 1,000,000 PJ, similar to that estimated for the eruption of the volcano Laki in Iceland in 1783.

Another terrestrial event is the earthquake. These occur rather frequently and also release vast amounts of energy. An earthquake having a magnitude of 8 on the Richter scale, which is large but occurs not infrequently, corresponds to an energy release of 50 PJ. This is the same amount as that released by the fall of the Tunguska object (Table 9.3).

9.4.3 The future terrestrial climate

As is known, the terrestrial climate is a rather unstable or, more exactly, metastable system (see Chapter 6). The number of parameters influencing climate seems to be very large. Some of the more important parameters are repeated in Table 9.4.

Had the Earth been situated slightly closer to the Sun, a runaway greenhouse effect would have occurred fairly early in the Earth's history. Had

Table 9.4. Impact of Selected Parameters on the Climate and Evolution, Partially According to S.H. Dole (1970) and M.H. Hart (1979)

Parameter	Present	Change	Results
Mass of Earth	1.0	×2	Surface gravity 1.38 times present, progression of life from sea to land slower, mountains lower, atmosphere denser.
		×1/2	Surface gravity of 0.73 times present, skeletons lighter, trees taller, mountains higher
Inclination of Earth's axis	23.5°	60°	Seasonal weather changes more extreme, only 5 degrees north or south of equator suitable for life, the rest too hot or too cold.
Distance, Sun–Earth	1.0	0.95	A runaway greenhouse effect would have occured early in Earth's history (Venus type of climate)
		1.01	Runaway glaciation would have occured about 2 billion years ago.
Earth's rotation	24 hrs	3 hrs	Oblateness would be pronounced, day/night temperature changes small
		100 hrs	Day/night temperature changes extreme, few life forms on continents
Mass of Sun (solar constant of present level – now "G2" class)	1.0	1.2	Period of rotation 1.54 years, a star of "F5" class, with life of $\sim 5.4 \times 10^9$ years; spectrum shifted to UV. "F5" is more active (eruptions)
		0.8	The year ~ 0.59 present year, a star of "G8" class with life of 20×10^9 years, spectrum shifted to IR
Distance, Earth–Moon	1.0	0.33	Tidal forces strong, rotation of Earths slower, day 6.9 times longer, continents uninhabitable
		~ 2	Loss of Moon, tides weak
Jupiter's mass	1.0	×1000	Second star? Earth's orbit unchanged?

the Earth's orbit been slightly larger instead, runaway glaciation would have occurred about 2 billion years ago. The continuously habitable zone around a solar-type star is, in fact, rather narrow, extending roughly from 0.95 to 1.01 astronomical units from the Sun.

A decrease of 1.6% in the solar constant, which would result if the Earth were only 0.8% further from the Sun, or the Sun's mass were 1.95×10^{30} kg instead of 1.98×10^{30} kg, would again cause runaway glaciation.

It seems the main unstable factor for the extropolated future is the terrestrial climate, but this need not lead to a catastrophe.

9.5 The Possibilities for Mankind: Self-destruction, Self-isolation, Expansion

The most dangerous event must be expected not from cosmic space nor from the inside of the Earth. The real danger for mankind is Man himself. A large nuclear war is much more probable and much more dangerous than all the other cosmic catastrophes.

In the open literature a picture of a nuclear world war at a level of 10^4 megatons has been discussed. This means the use of nuclear weapons having a total explosive power equivalent to 10,000 megatons of TNT (tri-nitro-toluene). Since the energy of 1 megaton of TNT equals 4 PJ (4×10^{15} J), a total world war of such magnitude would release an energy of 40,000 PJ.

It is perhaps incredible to note that all the known or possible cosmic catastrophes having an effect on the Earth up to today could have given a total radiation energy on the Earth's surface of approx. 1000 J/m². Nuclear war with 10^4 megatons, on the other hand, results in a mean energy distribution of 1,000,000 J/m² on only one-tenth of the Earth's surface. But the most dangerous consequence of a nuclear war is the long-lived radioactivity of fission products and the activated material of the soil.

Some interesting and rather unexpected data on cosmic and man-made catastrophes are collected in Figure 9.5. It must be clear from this that Man's greatest enemy is not the cosmos, not the violent Earth, but Man himself.

Another source of danger, this time not man-made, is built into the biosphere itself. Figure 9.6 gives an idea of these problems. The genetic, somatic, or mental degeneration of mankind cannot be ruled out. Nevertheless, there are no grounds for fearing such a danger. On the contrary, there is also a basis for a rather optimistic view of our own future here.

Mankind, in spite of being part of the biosphere, has its own logic of evolution. From the point of view of this chapter the most important parameters, such as population increase and cosmic activity, are shown in Figure 9.7.

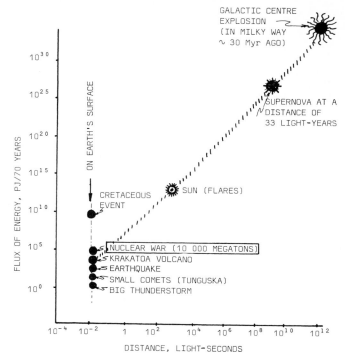

Figure 9.5. Catastrophic events on the Earth.

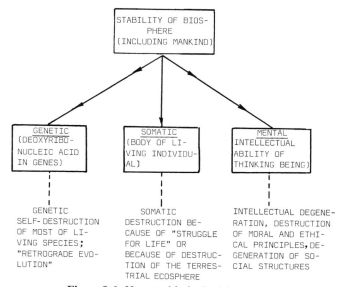

Figure 9.6. How stable is the biosphere?

9.6 Human Colonies in Space – Possibility or Nonsense?

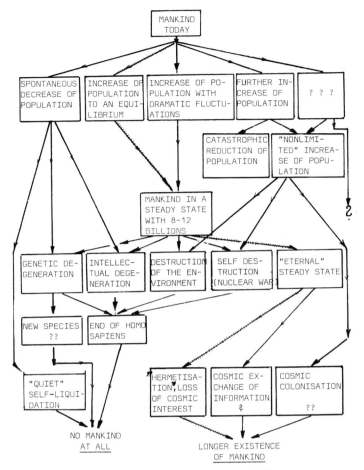

Figure 9.7. The future of mankind.

Here only some options are shown. These are still open and which way mankind will choose cannot be predicted on the basis of present-day knowledge. Perhaps this question will always be open with no objective way of finding an answer. Man himself is the biggest unknown.

9.6 Human Colonies in Space – Possibility or Nonsense?

In spite of many possible pessimistic or dangerous prognoses, an attempt is made here to discuss the future of mankind in a rather optimistic way.

It is perhaps surprising to see the large number of books and papers, the large amount of academic activity and the specialist groups all dealing with the idea of setting up colonies in space.

The term "space colonies" includes the following:

- man-made spaceships on a huge scale containing thousands or even tens of thousands of people;
- such spaceships after a development phase becoming self-sustaining and fully independent of the Earth and of the planetary civilisation;
- spaceship remaining in the neighbourhood of the Earth or at least of the Sun.

The postulated scenarios nearly all contain the following aspects:

- The space colonies would be built of lunar material. This gives a significant improvement in energy requirements for material transport from the Earth to the colony.
- After an early development phase, during which workers would be housed in modular habitation, the first space colony might be expected to be a sphere ~ 500 m in diameter housing up to 10,000 people. It would rotate twice per minute to give a sensation of gravity equivalent to that on Earth, at the equator of the colony. Towards the "poles" the "gravity effect" would decrease.
- Giant mirrors would reflect the sunlight through "windows".
- The colony would be fully independent.

Why is this a popular proposal? One reason for a space colony would be in case of impending catastrophe on the Earth which could wipe out civilisation. The colonies would then have to survive on their own.

Colonies have been proposed in various orbits around the Sun, even beyond the orbit of Pluto. However, further than 3 light-days away the intensity of sunlight on solar panels would become too low to maintain a comfortable climate in the colonies.

Now I wish to present a personal opinion about these proposals and the related discussions on space colonies: they are a waste of time and get in the way of proper consideration of the real problems facing mankind.

9.7 The Existence of Other Planetary Systems with Intelligent Life

9.7.1 How many stars have planetary systems?

An early dream of mankind was, perhaps, to colonise other planets. Today it is realised that this dream must involve not the other planets of the Solar System but rather other planetary systems. The colonisation of the neighbouring planets, Venus and Mars or the Moon, with what we now known of their

9.7 The Existence of Other Planetary Systems with Intelligent Life

characteristics, seems to be accepted as hardly possible. The dream remains, however, of conquering other planets, those lying outside the Solar System.

What is known today about other planetary systems?

First a short discussion of the possible existence of habitable planets in this Galaxy, which means mainly towards the centre of the Galaxy where the majority of the stars are clustered.

The Galactic centre contains approximately 30,000 stars per cubic light-year, which gives an average distance between stars of approx. 0.05 light-years. This is a million times greater than the star density in the neighbourhood of the Sun, which is one star in 400 cubic light-years (average distance 7 light-years). Such a density means that any being on a planet circling a star near the centre of the Galaxy would see a million stars as bright as Sirius, the brightest star in our sky. (With the naked eye one can see only a few thousand stars from Earth and most of these are very faint.) The integrated intensity of all the stars in the night sky of such a planet in the Galactic centre would equal that of 200 full Moons. With such a high stellar density close encounters between stars would be so frequent that planets would be ripped out of their orbits every few hundred-million years, therefore

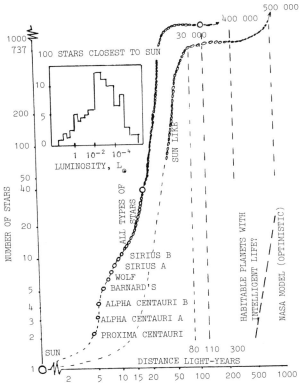

Figure 9.8. Expected number of habitable planets.

stable planetary systems seem to be impossible there. It is known that at a radius of 20 light-years from our Sun there are approx. 100 stars. The average star density is 0.01 star per cubic light-year – a close encounter is improbable.

The only known planets in the whole Universe are the nine in the Solar System. It is not certain whether other stars have planets and there are no reliable instruments for detecting planets around neighbouring stars with certainty. Some proposals have been made for a spinning infrared interferometer which might be capable of discovering the existence of planets within a range of 30 light-years (Figure 9.8).

This shows the difficulties involved in detection of the presence of planetary systems from the Earth: the alteration in the motion of the Sun (~ 1 microsecond of arc) from a distance of 30 light-years by an astronomer is the same as the observation of an atom in your finger when viewed at arm's length.

As long as our knowledge of the real origins of our planetary system (so-called "cosmogony") remains weak and the observational possibilities remain limited, rigorous discussion of the possibility of extraterrestrial life or extraterrestrial intelligence is impossible.

9.7.2 How many planets having intelligent life could exist?

A number of questions immediately arise, some of them currently unanswerable:

- What is life?
- What is intelligent life?
- Is there intelligent life on the Earth?
- Does Man have the ability to discover other life forms, particularly intelligent life?
- If he does have this ability, what is the most likely property of intelligent life one could search for at large distances?
- Is the emission of coherent electromagnetic waves (radio emissions) the best (or, in fact, the only) presently feasible method?
- Can one assume that the classification of a civilisation is best made according to the amount of energy it consumes, marking the level of its development?

One should recall the quotation of S. Lem at the beginning of this chapter.

There are other collections of questions connected with the probability of finding extraterrestrial civilisations which have been discussed for at least the last decade.

Naturally there are many criticisms of the type of calculation in Figure 9.9, but it must be realised that no better alternative exists at the present time.

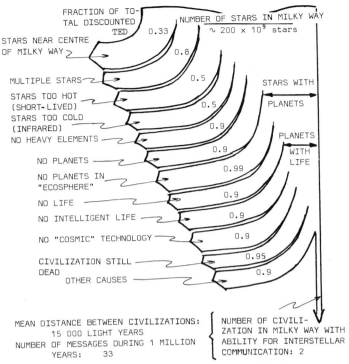

Figure 9.9. How to estimate the number of civilisations.

9.8 The Extraterrestrial Exchange of Information

Let us assume that there is a possibility of contact with an extraterrestrial civilisation. So long as one accepts the present state of the natural sciences, one must acknowledge the fact that the quickest method of transporting information is by using the massless particles as carriers.

The result of such consideration is, of course, obvious. The best medium for carrying information in cosmic space is coherent radiation in the form of electromagnetic waves or photon emission, in other words, radio waves (Figures 9.10 and 9.11).

The question must be asked, what is the aim of the search for extraterrestrial intelligence? What can be gained? Can a cost–benefit analysis be made?

The answer could be given in different ways:

- technical: the hope of finding support for man's technical civilisation – new technical ideas, new information;
- scientific: a purely heuristic aim: to enlarge man's knowledge of the Universe and the natural world;

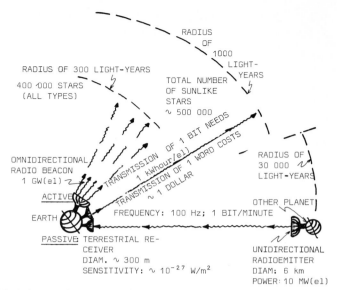

Figure 9.10. Scheme of interstellar communication. The number of possible variables is very large (e.g.: direction, frequency, band width, integral time, polarisation, duty cycle): total 10^{26} variables; still used number of parameters: 1.5×10^7 variables.

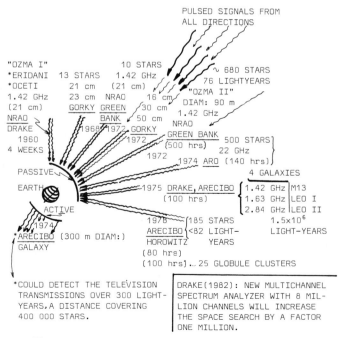

Figure 9.11. The search for extraterrestrial intelligence.

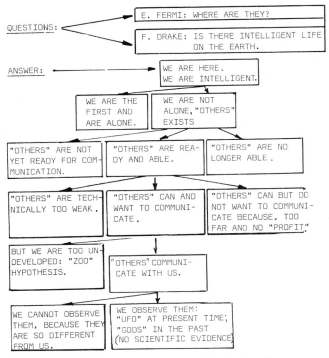

Figure 9.12. Ourselves and the "others".

- religious: ...
- moral: if other civilisations have existed longer than mankind this will give us hope that mankind also has a future.

The reader himself must find the reason why the search for extraterrestrial civilisation is, for him, worthwhile. In so doing he should not forget the real costs of this activity, which may be at the expense of other more worthwhile tasks here on Earth.

One should also remember that the others may not be not interested in contacting us (Figure 9.12).

9.9 Summary of the Limits of World Population Growth

The cosmic activity of mankind can be classified, from the point of view of strategic aims, into three groups:

- the multiplication of knowledge of the Universe, a purely heuristic aim;
- the search for information exchange with extraterrestrial civilisations: technical, scientific, or even moral and ethical questions;
- the discovery of "habitable" planets as a target for future colonisation.

The last point is considered in the next section – the possibility of cosmic colonisation.

The first question: why are plans needed for expansion? Is this due to the increasing pressure of population?

9.10 Human Galactic Expansion and the Drake Limit

9.10.1 The expansion velocity

How often has the future expansion of mankind far from the Earth, deep in galactic space, been discussed and how often has it been said that this is the real solution to all of Man's troubles? The brave new cosmic world.

But some rather competent people have another opinion on the distant cosmic expansion of mankind.

S. Van Hoerner, an expert in this area, has said the following:

"Even with perfect technology the problem (of population explosion) cannot be solved by interstellar expansion. Frank Drake once suggested over [a] coffee break that the finite speed of light sets a limit and this turns out to be right. If we populate all habitable planets within a sphere of increasing radius such that the volume of the sphere increases at 2%/year (our present growth rate), then the limit is reached when the radius of the sphere increases at the speed of light. The resulting numbers are amazingly small. The limiting radius is only 150 light-years. Within this sphere are 30,000 habitable planets but, starting with 1 today and increasing 2 percent per year, it takes only 500 years to populate all of them with the same density as the Earth has now".

An explanation of this calculation is given here:

– The Earth's 4 billion people decide that this figure is not to be exceeded on Earth.
– The annual rate of increase of 2%, that is, 80 million extra people per year, must be exported to the neighbouring planetary system.
– The cosmic emigration must be realised by means of super-rockets having the following parameters: velocity of flight, 1/10th the speed of light; number of passengers, 1000.
– Since the other planets in the Solar System are not hospitable it is decided to colonise the neighbouring star system. It has been assumed the other stars, or even groups of double or triple stars, independent of their luminosity and surface temperature, have at least one planet which is acceptable for human life and has room to accept approx. the same number of people as on Earth, that is, 4 billion.
– The average distance of neighbouring stars (not only sun-like) is taken as 7 light-years (see Figure 9.8). In a radius of 15 light-years there are 40 stars; in

9.10 Human Galactic Expansion and the Drake Limit

a radius of 110 light-years, 30,000 stars; in a radius of 50,000 light-years, 100 billion stars.
- The cosmic colonisation begins at once.
- The first super-rockets have to travel for approx. 70 years (distance 7 light-years with a velocity of 1/10 the speed of light, 30,000 km/s).
- At the end of the flight the rocket can contain only 1000 passengers. This means it can start with only 250 passengers allowing for normal reproduction during the 70 years of flight at a 2% increase per annum.
- The rate at which the rockets must leave the Earth, each with 250 passengers, is 36 every hour, which corresponds to the emigration of 80 million people per year.
- After 70 years from the start of the era of cosmic colonisation about 5.6 billion colonists have been dispatched, and in all the rockets then under way there are approx. 12 billion persons. This is enough to fill three planets to capacity. At the same moment there are 22 million rockets under way but only 12 million can land on the next three planets. The rest must look for more distant stars and therefore must increase their velocity above 1/10 the speed of light. If this is not done the flight will be too long and the rocket will become overcrowded.
- The rockets, then having started from Earth, must also have a larger velocity if they are to reach still empty planets at a greater distance than those attempted by the rockets then in mid-flight.
- 520 years after the start of the whole exercise the size of the expanding cosmic population will be 120 thousand-billion! This is equivalent to 30,000 full planets each with 4 billion people.
- These 30,000 planets are all within a distance of 110 light years from the Earth. They are all fully occupied within 520 years.
- 35 years after that point – the next doubling period (2% per annum gives a doubling time of 35 years) – the cosmic population is 240 thousand-billion and a further 30,000 planets are needed.
- Given the average density of stars in the Milky Way, these additional 30,000 planets are contained in a spherical shell with an inner radius of 110 light-years and an outer radius of 145 light-years – that is, 35 light-years thick.
- Repeating the situation once more: after 520 years from the start of cosmic colonisation, and over a period of 35–555 years from the start, the cosmic population increases by another 120 thousand-billion people. This extra population can only be housed on the planets lying in a shell 35 light-years thick (from 110–145 light-years from Earth).
- From this it is clear that the cosmic population has to spread out into space at the speed of light (35 light-years in 35 years!). Even then the subsequent colonists have no further ways of colonisation.

There is no doubt that decreasing the rate of population growth from 2% to only 1% per annum cannot change the situation significantly. It only makes

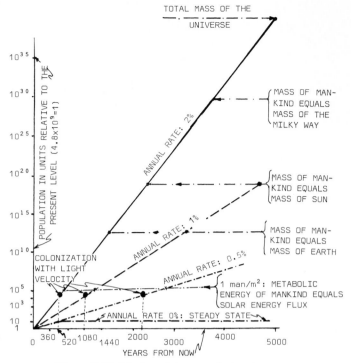

Figure 9.13. Limits of the population increase.

the limiting case twice as far off. Instead of 520 years later, the crisis of the colonisation at the speed of light will occur in 1040 years time (Figure 9.13).

9.10.2 The energy need for cosmic journeys

After this calculation it is perhaps pointless to criticise cosmic colonisation from the point of view of technical feasibility. Perhaps, however, for some "technocrats" this will be a stronger argument than that given above. Figure 9.14 and Table 9.5 are probably sufficient for this purpose.

In the late 1970s the results of an English project named *Daedalus* were published. This concerned a trip to the next potential planetary system called "Barnard's star" at a distance of 6 light-years. The rocket would be 230 m long, weighing 50,000 tons, propelled by the energy of nuclear fusion of deuterium and helium-3 ignited by electron guns. Since helium-3 is so scarce on Earth, the rocket would have to take fuel from the abundant gaseous atmosphere of Jupiter.

The spacecraft would be under power for nearly four years, reaching a top speed of 130×10^6 km/hr, that is, 12% of the speed of light. The total trip would last 50 years. It is proposed that "Daedalus" be built in the next century from industrial bases on the Moon.

9.10 Human Galactic Expansion and the Drake Limit

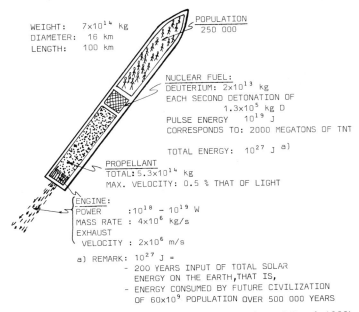

Figure 9.14. Hypothetical "world ship" (Martin and Bond, 1980).

Table 9.5. A Two-Way Trip to Alpha-Centauri (~4.5 light-years each way)

Mass ratio	$\mu = \dfrac{M_i}{M_o} =$	$\dfrac{\text{initial total mass of rocket + fuel}}{\text{mass of rocket after burnout}}$
Fraction of fuel converted into energy	$\varepsilon =$	$\dfrac{\text{energy release from fuel } (E/c^2)}{\text{mass of fuel}}$
Velocity of rocket	$v =$	$\dfrac{\text{velocity of rocket}}{\text{velocity of light}}$

	Uranium-235 fission reactor $\varepsilon = 7 \times 10^{-4}$	Deuterium fusion reactor $\varepsilon = 40 \times 10^{-4}$
v	μ	μ
0.1	3.8×10^4	8.1×10^1
0.2	2.3×10^9	6.2×10^3
0.3	not feasible	1.1×10^6
0.4	not feasible	1.5×10^8

Just sending one spaceship to a nearby star and returning it at one tenth of the speed of light using neither fusion nor fission drive (both having relatively low yields), but instead using the most effective energy source – matter-antimatter annihilation – would require 10^{24} J. The solar energy flux absorbed by the Earth is 1.2×10^{17} J/s, and this corresponds to 3 months' sunshine or 3000 years of energy as presently consumed by mankind.

If one seeks to reduce the energy requirement of space travel by, say, a factor of 10,000 (that is, to a velocity of 300 km/s), then the round trip requires a travel time of 400 years/light-year, or 1600 years to the nearest star.

A different type of spacecraft has also been proposed, the cosmic ramjet.

The ramjet is designed to utilise the burning of deuterium in the interstellar medium. If the interstellar plasma is optically thick, then, for minimum plasma compression, one requires an energy generation close to the solar luminosity and a ram scoop 1/2 light-year in diameter. If the plasma is optically thin, then its radioactive emission exceeds the energy gain from thermonuclear burning by at least nine orders of magnitude. It is thus concluded that the interstellar ramjet is an unlikely propulsion mode.

9.11 Is It Really Impossible to Colonise the Galaxies?

In many considerations of extraterrestrial intelligence, the idea of a civilisation which has conquered a whole Galaxy has been examined. From the previous calculations this seems to be impossible, but only at first glance. A deeper discussion shows that the conquest of a Galaxy by one civilisation would not be impossible.

Here is a model of such a conquest:

- velocity of travel 1/10 the speed of light;
- after the conquest of a new planet, an arbitrary period of 500 years is allowed for full regeneration and assimilation before the next colony is founded;
- this gives an effective expansion velocity of 0.016 the speed of light;
- the Galaxy has a diameter of 100,000 light-years with 100 billion stars, each with one habitable planet (an extremely optimistic assumption);
- to colonise the whole Galaxy the following period of time is needed:

$$\frac{100{,}000 \text{ light-years}}{0.016 \text{ light speed}} = 5 \times 10^6 \text{ years};$$

- after 5 million years all the 100 billion star systems, each with a habitable planet, are colonised;
- this 5 million years corresponds to the following number of doubling

periods: for 10^{11} habitable planets,

$$2^{36} = 10^{11},$$

that is, 36 periods each of

$$\frac{5 \times 10^6 \text{ years}}{36 \text{ periods}} = 140{,}000 \text{ years};$$

– for a doubling period of 140,000 years, the annual rate of increase must be:

$$0.00052\% \; (5.2 \times 10^{-6}) \text{ per year.}$$

This is an interesting result. The conquest of the Galaxy by one civilisation is possible but needs a great deal of time – at least 5 million years. After this period, if all of the stars have one habitable planet each, then the Galactic population is 100 billion times greater than the terrestrial population:

$$10^{11} \times (4 \times 10^9) = 4 \times 10^{20} \text{ people.}$$

This figure is approximately the same as the number of stars in the Universe.

The most significant result is the value of the permitted annual rate of population increase. It should not be greater than 5.2×10^{-6} per year. Only under these conditions could overcrowding be avoided during the expansion.

One must ask why mankind must follow such a rate of increase of exactly 0.00052%? Why not 0.01%? The answer is unequivocal, because otherwise the population will increase so fast that it will come up against the limit of expansion given by the speed of light.

Why not a 0.00% annual increase? In this case there is no interest in cosmic expansion at all. The Earth will satisfy our needs.

9.12 The Very Distant Future; Mankind on This Planet

This section is left to the reader's imagination.

9.13 What Are the Conclusions Concerning the Distant Future of Mankind?

From what was discussed in this chapter:

– the stability of the Universe, Galaxy, and Sun;
– the continuing evolution of the Earth, its atmosphere, hydrosphere, and lithosphere;

- the probable evolution of the biosphere and the biological evolution of Man;
- the chance of avoiding self-destruction and self-degeneration;
- the possibility of human colonies in space;
- the possibility of the colonisation of neighbouring planets and even the entire Galaxy;
- the likelihood of extraterrestial exchange of information;

the following conclusions can be drawn for mankind's future developement:

- It seems possible that Man can exist, at least under stable population conditions, for a long period on this planet, even billions of years.
- The search for cosmic solutions, that is, cosmic expansion, is not sufficiently justified.
- The real solutions to Man's future problems lie in self-limitation of population and development "in depth".

Finally, and emphatically, it must be repeated that Man's future is in his own hands.

Bibliography

Chapter 1

Carr, B.J. and Rees, M.J. The anthropic principle and the structure of the physical world. *Nature* **278**, 403 (1979).
Carter, B. The anthropic principle and its implications for biological evolution. *Phil. Trans. R. Soc. Lond.* **A310**, 347 (1983).
Davies, P.C.W. *The Accidental Universe*. Cambridge: Cambridge Univ. Press, 1982.
Glashow, S.L. Towards a unified theory. *Rev. Modern Phys.* **52**, 539 (1980).
Rees, M.J. Large numbers and ratios in astrophysics and cosmology. *Phil. Trans. R. Soc. Lond.* **A310**, 311 (1983).
Salam, A. Gauge unification of fundamental forces. *Rev. Modern Phys.* **52**, 525 (1980).
Taube, M. An empirical formula for the coupling constants of the four elementary interactions. *Atomkernenergie* **40** (2), 128 (1982).
Taube, M. A formula for the calculations of the ratios of the masses of elementary particles. *Atomkernenergie* **40** (3), 208 (1982).

Chapter 2

Dolgov, A.D., and Zeldowich, Ya.B. Cosmology and elementary particles. *Rev. Mod. Phys.* **53**, 1 (1981).
Guth, A.H., and Steinhardt, P.J. The inflationary Universe. *Sci. Am.* **250**, 90 (1984).
Harrison, E.R. *Cosmology*. Cambridge: Cambridge Univ. Press, 1981.
Novikov, I.D. *Evolution of the Universe*. Cambridge: Cambridge Univ. Press, 1983.
Olive, K.A., Schramm, D.N., Steigman, G., Turner, M.S., and Kang, J. Big bang nucleosynthesis as a probe of cosmology. *Astrophys. J.* **246**, 557 (1981).
Peebles, P.J.E. The origin of galaxies and clusters of galaxies. *Science* **224**, 1385 (1984).
Rees, M.J. Our Universe and others. *Quat. J. Roy. Astr. Soc.* **22**, 109 (1981).
Van den Bergh, S. Size and age of the Universe. *Science* **213**, 825 (1981).
Weinberg, S. *The First Three Minutes*. New York: Basic Books Inc., 1977.
Weisskopf, V.F. The origin of the Universe. *Am. Sci.* **473**, 71 (1983).

Chapter 3

Audouze, J., and Vauclair, S. *An Introduction to Nuclear Astrophysics*. Dordrecht: Reidel, 1980.
Blitz, L., Fich, M., and Kulkarvi, S. The new Milky Way. *Science* **220**, 1233 (1983).

Burbidge, E., Burbidge, G., Fowler, W., and Hoyle, F. Synthesis of elements in stars. *Rev. Modern Physics* **29** (4), 458–650 (1957).
Gehrz, R.D., Black, D.C. and Solomon, P.M. The formation of stellar systems from interstellar molecular clouds. *Science* **224**, 823 (1984).
Gibbons, G.W., and Hawkins, S.W. (Eds.) *The Very Early Universe.* Cambridge: Cambridge Univ. Press, 1983.
Kourganoff, V. *Introduction to Advanced Astrophysics.* Dordrecht: Reidel, 1980.
Meier, D.L., and Sunayev, R.A. Primeval galaxies. *Sci. Am.* **241** (5), 106 (1979).
Sato, H., and Takamara, F. Formation of galaxies in the neutrino dominated Universe. *Prog. Theor. Phys.* **66**, 508 (1981).
Shkolvskii, J.S. *Stars, Their Birth, Life and Death.* San Francisco: Freeman, 1978.
Tinsley, B.M., and Larson, R.B. (Eds.). *The Evolution of Galaxies and Stellar Population.* New Haven: Yale Univ. Observatory, 1977.
Turner, M.S., and Schramm, D.N. Cosmology and elementary-particle physics. *Phys. Today* **9**, 42 (1979).
Weinberg, S. *The First Three Minutes: A Modern View of the Origin of the Universe.* New York: Basic Books Inc., 1977.
Whitmire, D.P., and Jackson, A.A. Are periodic mass extinctions driven by a distant solar companion? *Nature*, **308**, 713 (1984).

Chapter 4

Bury, J.S. The planet Venus. *J. Brit. Interplan. Soc.* **32**, 123 (1979).
Davis, M., Hut, P., and Muller, R.A. Extinction of species by periodic comet showers. *Nature* **308**, 715 (1984).
Eigen, M. Self-organisation of matter and evolution of biological macro-molecules. *Naturwissensch.* **10**, 456 (1971).
Gehrels, T. (Ed). *Protostars and Planets.* Tucson: Univ. Arizona, 1978.
Goodwin, A.M. Precambrian perspectives. *Science* **213**, 55 (1981).
Green, S. Interstellar chemistry. *Ann. Rev. Phys. Chem.* **32**, 103 (1981).
Hart, M.H. The evolution of the Earth. *Icarus* **33**, 23 (1978).
Margulis, L. *Symbiosis in Cell Evolution.* San Francisco: Freeman, 1981.
Martin, P.G. *Cosmic Dust.* Oxford: Claredon Press, 1978.
McElhinny, M.W. *The Earth: Its Origin, Structure, and Evolution.* London: Academic Press, 1979.
NASA, *First Symposium on Chemical Evolution and the Origin and Evolution of Life.* D.L. De Vincenzi and L.G. Pleasant (Eds.). NASA, Conf. Publ. 2276, 1983.
Orgel, L.E. *The Origins of Life.* London: Chapman Hall, 1973.
Ponnamperuma, C. (Ed). *Chemical Evolution of the Giant Planets.* New York: Academic Press, 1976.
Press, F., and Siever, R. *Earth.* San Francisco: Freeman, 1982.
Press, W.H., and Lightman, A.P. Dependence of macrophysical phenomena on the values of the fundamental constants. *Phil. Trans. R. Soc. Lond.* **A310**, 323 (1984).
Sagan, C., Cameron, A.G.W., Parker, E.N., Murray, B.C., Siever, R., Wood, J.A., Pollack, J.B., Wolfe, J.H., Hunten, D.M., Hartmann, W.K., and Allen, J.A. The solar system. *Sci. Am.* **233**, 3 (1975).
Scoville, N., and Young, M.S., Molecular clouds star formation and galactic structure. *Sci. Am.* **250**(4), 30 (1984).
Siever, R., Jeanloz, R., Mc Kenzie, D.P., Francheteau, J., Burchfiel, B.C., Broecker, W.S., Ingersoll, A.P., Cloud, P., The dynamic Earth. *Sci. Am.* **249**, 3 (1983).
Taube, M. *Hydrogen as Carrier of Life*, Warsaw: Nucl. Publ. House, 1966.

Taube, M. et al. Synthesis of amino-acids and their precursors at high temperature. *Angewandte Chemie* **79**, 239 (1967).
Taube, M. Simple model of a living thing. *First Europ. Biophysics Congress, Baden, Vienna 1071.* Verlag Wiener Medizin, Akad. 1971.
Taube, M. Chemie in interstellarem Raum. *Chimia* **4**, 131 (1977).
Tingsley, B.M., and Cameron, A.G.W. Possible influence of comets on the chemical evolution of the galaxy. *Astroph. Space Sci.* **31**, 31 (1974).
Vidal, G., The oldest eukaryotic cells. *Sci. Am.* **250** (2), 32 (1984).
Wald, G. Fitness in the Universe; choices and necessities. *Origin of Life* **5**, 7 (1974).
Windley, B.F. *The Evolving Continents.* London: J. Wiley, 1977.
Wynn-Williams, C.C. (Ed). *Infrared Astronomy.* Dordrecht: Reidel, 1981.

Chapter 5

Budyko, M.I. *Global Ecology.* Progress Pub., Moscow, 1980.
Chapman Conference, *Natural variations in carbon dioxied in carbon cycle.* Tarpon Springs, USA, 1984.
Ehrlich, P.R., Ehrlich, A.H., and Holdren, J.P. *Ecoscience: Population. Resources, Environment.* Freeman, San Francisco, 1977.
Francis, P., and Self, S. The eruption of Krakatau. *Sci. Am.* **249**(5), 146 (1983).
Gabor, D. *Beyond the Age of Waste.* Oxford: Pergamon Press, 1981.
Garrels, R.M., Mackenzie, F.T., and Hunt, C. *Chemical Cycles and the Global Environment.* Los Altos: Kaufman, 1975.
Global 2000, *Report to the President.* US Government Printing Office, Washington, 1980.
Goeller, H.E., and Weinberg, A.M. The age of substituability. *Science* **191**, 683 (1976).
Holland, H.D. *The Chemistry of the Atmosphere and Oceans.* Wiley, New York, 1978.
Holser, W.T. Catastrophic chemical events in the history of the ocean. *Nature,* **267**, 403 (1977)
Kalenin, G.P., and Bykov, V.D. The world's water resources, present and future, in *Ecology of Man,* R.L. Smith (ed.), Harper, New York, 1972.
Leontief, W.W., Carter, A.P., and Petrie, P. *The Future of the World Economy.* United Nation Study, Oxford Univ. Press, New York, 1977.
Maddox, J. From Santorini to Armageddon. *Science* **307**, 107 (1984).
National Academy of Science, *Causes and Effects of Changes of Stratospheric Ozone, Update 1983.* N.A. Press, Washington, 1984.
Seiler, W., and Crutze, P.J. Estimates of gross and net fluxes of carbon between the biosphere and the atmosphere. *Climatic Change* **2**, 207 (1980).
Siever, R., Jeanloz, R., McKenzie, D.P., Francheteau, J., Burchfiel, B.C., Broecker, W.S., Ingersoll, A.P., and Cloud, P. The dynamic Earth. *Am. Sci.* **249**(3), 30–145 (1983).
Skinner, B.J. Earth resources. *Proc. Nat. Acad. Sci. USA* **76**, 4212 (1979).
Walsh, J. et al. Biological export of shelf carbon is a sink of the global CO_2 cycle. *Nature* **291**, 196 (1981).
World Meteorological Organisation, *Analysis and interpretation of atmospheric CO_2 data,* Bern, Sept. 1981. *Nature* **295**, 190 (1982).

Chapter 6

Bergmann, K.H., Hecht, A.D., and Schneider, S.H., Climate models. *Phys. Tod.* **10,** 44 (1981).
Budyko, M.I., *Climatic Changes.* Am. Geog. Un., Washington, 1977.
Covey, C., The Earth's orbit and the ice ages. *Sci. Am.* **250** (2), 42 (1984).
Deffeys, K.S., and Mac Gregor, I.O., World uranium resources. *Sci. Amer.* **242 (1),** 50 (1980).
De Meo E.A., and Taylor R.W. Solar photovoltaic power systems. *Science* **224,** 245 (1984).
Energy In a Finite World, Mc Donald, A., Ed., Int. Inst. Appl. Syst. Anal., Luxembourg, 1981.
Fettweis, G.B., *World Coal Resources.* Elsevier, Amsterdam, 1979.
Glasstone, S., and Jordan, W.H., *Nuclear power and its environmental effects.* Am. Nucl. Soc. Lagrange Park, Illinois, USA, 1980.
Hallam, A., The causes of mass extinctions. *Nature* **308,** 686 (1984).
Hansen, J. et al., Climatic impact of increasing atmospheric carbon dioxide. *Science* (213), 957 (1981).
Hart, M.H., The evolution of the atmosphere of the Earth. *Icarus* **33,** 23 (1978).
Isaacs, J.D., and Schmitt, W.R., Ocean energy: forms and prospects. *Science* **207,** 265 (1980).
Kellog, W.W., Influences of mankind on climate. *Ann. Rev. Earth Plan. Sci.* **63,** (1979).
Kulcinski, G.L., Kessler, G., Holdren, J., and Haefele, W., Energy for the long run: fission or fusion. *Amer. Sci.* **67,** 78 (1979).
Lamb, H.H., *Climate: Present, Past and Future,* Vols. 1, 2. Methuen, London, 1977.
Landsberg, H.H. (Ed.), World Survey of Climatology. Vol. 1: *General Climatology* (1982); Vol. 2: *General Climatology* (1969); Vol. 3: *General Climatology* (1981). Elsevier, Amsterdam.
Oeschger, H. et al., *Das Klima.* Springer-Verlag, Berlin-Heidelberg, 1980.
Pearson, R., *Climate and Evolution.* Academic Press, London, 1978.
Peixóto J.P., and Oort A.H. Physics of climate. *Rev. Mod. Phys.* **56** (3), 365 (1984).
Rudloff, W., *World Climates with Tables of Climatic Data and Practical Suggestions.* Wissensch. Verlagsgesel. Stuttgart, 1981.
Sagan, C., Toon, O.B., and Pollack, J.B., Anthropogenic albedo changes and the Earth's climate. *Science* **206,** 1363 (1979).
Schneider, S.H., and Thompson, S.L., Cosmic conclusions from climatic models. *Icarus* **41,** 456 (1980).
Taube, M., Improved inherent safety in liquid fuel reactors. *Atomkernenergie* **40,** 73 (1982).
Taube, M., *Plutonium.* Chemie Verlag, Weinheim, 1974.
Taube, M. et al., A system of hydrogen vehicles with liquid organic hydrides. *Int. Journ. Hydr. Energy* **8** (3), 213 (1983).
Upton, C.A., Biological effects of low level ionizing radiation. *Sci. Am.* **246** (2), 29 (1982).
Verhoogen, J., *Energetics of the Earth.* Nat. Acad. Sci. USA, Washington, 1980.
Wallace, R.H. et al., *Assessment of geopressurised-geothermal resources.* US Geolog. Survey. Circ. 790 (1979).
World Climate Conference. World Meteorological Organisation, Geneva, February 1979.
XII. World Energy Conference, Proceedings. New Dehli, September 1983.

Chapter 7

Allbaby, M., *World Food Resources*. Appl. Science Pub., London, 1977.
Dritschilo, W. et al., Energy is food resource ratios for alternative energy technologies. *Energy* **4**, 255 (1983).
Duckham, A.N. et al. (Eds.), *Food Production and Consumption*. North Holland, Amsterdam, 1976.
Jensen, N.F., Limits to growth in world food production. *Science* **201**, 317 (1978).
Kerr, R.A., Periodic impact and extinctions. *Science*, 1277 (1984).
Leach, G., *Energy and Food Production*. IPC Sci. Tech. Press, London, 1976.
Maddox, J., Extinction by catastrophe. *Nature* **308**, 685 (1984).
Nitecki, M.H. (Ed), *Biotic Crisis in Ecological and Evolutionary Time*. Academic Press, New York, 1981.
Pirie, N.W., Waste not, want not. *New Scientist* **75**, 233 (1977).
Revelle, R., Food and population. *Sci. Am.* **231** (3), 160 (1974).
Russell, D.A., Mass extinctions of the late Mesozoic. *Sci. Am.* **246**, 48 (1982).
Spurr, S.H., Silviculture. *Sci. Am.* **240**(2), 62 (1979).
Pilbeam, D., The descent of hominoids and hominids. *Sci. Am.* **250** (3), 60 (1984).
Woodwell, G.M. et al., Global deforestation: contribution to atmospheric carbon dioxide. *Science* **222**, 4628, 1081 (1983).

Chapter 8

Bronowski, J., *The Ascent of Man*. BBC, London, 1973.
Ehrlich, P.R., Ehrlich, A.H., and Holdren, J.P., *Ecoscience, Population, Resources, Environment*. Freeman, San Francisco, 1977.
Gilland, B., *The Next Seventy Years*. Abacus Press, Kent, 1979.
Global 2000, *Report to the President*. US Govt. Printing Office, Washington, 1980.
Interfutures, *Facing the Future*. OECD, Paris, 1979.
Kahn, H., Brown, W., and Martel, L., *The Next 200 Years*. William Morrow, New York, 1976.
Keyfitz, N., The population of China. *Sci. Am.* **250**(2), 22 (1984).
Meadows, D.L., *The Limits of Growth*. DVA, Stuttgart, 1972.
Randers, J., and Zahn, E.K.O., *Dynamics of Growth in a Finite World*. Wright-Allen Press, Cambridge, 1974.
Skinner, B.J., Earth Resources. *Proc. Natl. Acad. Sci. USA*, **76**(9), 4112 (1979).
Taube, M., Future of the terrestrial civilization over a period of billions of years. *Journ. Br. Interplanet. Soc.* **35**, 219 (1981).
WEC, *World Energy Demand to 2020*. IPC Sci. Tech. Press. New York, 1978.

Chapter 9

Asimov, I., *A Choice of Catastrophes*. Fawcett Columbine, New York, 1981.
Black, D.C., In search of other planetary systems. *Space Sci. Rev.* **25**, 35 (1980).
Bond, A., and Martin, A.R., A conservative estimate of the number of habitable planets. *J. Brit. Interplan. Soc.* **33**, 101 (1980).
Clark, D.H., McCrea, W.H., and Stephenson, F.R., Celestial chaos and terrestrial catastrophes. *Nature* **265**, 318 (1978).

Davis, M., Hut, P., Muller, R.A., Extinction of species by periodic comet showers. *Nature* **308,** 715 (1984).
Dole, S.H., *Habitable Planets for Man*. American Elsevier, New York, 1970 (2nd Edit).
Dyson, F.J., Time without end; physics and biology in an open Universe. *Rev. Moderns Phys.* **51** (3), 447 (1979).
Eccles, J.C., Evolution of the brain. *Ann. New York Acad. Sci.* **299,** 161 (1977).
Hart, M.H., Habitable zones about Main Sequence stars. *Icarus* **37,** 351 (1979).
Horowitz, P., A search for signals of extraterrestrial origin. *Science* **201,** 733 (1978).
Islam, J.N., The long term future of Universe. *Vistas Astron.* **23** (3), 265 (1979).
Martin, A.R. (ed). "Project Daedalus". *J. Brit. Interplan. Soc. Suppl.*, (1978).
Martin, A.R., World ships, concept, cause, cost, construction and colonisation. *J. Brit. Interplan. Soc.* **37,** 243 (1984).
Matloff, G.L., and Mallove, E.F., The first interstellar colonization mission. *J. Brit. Interplan. Soc.* **33,** 84 (1980).
Morris, P., Ed. *The Search for Extraterrestrial Intelligence*. NASA, SP-419 (1977).
Papagiannis, M.D. (ed.), *Strategies for the Search for Life in the Universe*. Reidel Pub. Co., Dordrecht, 1980.
Pollard, W.G., The prevalence of Earthlike planets. *Am. Scientist* **67,** 653 (1979).
Shkolvskii, I.S., and Sagan, C., *Intelligent Life in the Universe*. Holden-Day, San Francisco, 1966.
Tang, T.B., The number of inhabited planets in the Galaxy. *J. Brit. Interplan. Soc.* **37,** 410 (1984).
Taube, M., Future of terrestrial civilization over a period of billions of years (Red Giant and Earth shift). *Journ. Brit. Interplan. Soc.* **35,** 219 (1982).
Taube, M., *Human Intelligence, the Ultimate Resource*. Swiss Fed. Inst. Technology, Zürich, 1983.
von Hoerner, S., Where is everybody? *Naturwiss.* **65,** 553 (1978).
Wolfendale, A., Cosmic rays and ancient catastrophes. *New Scientist* **634** (1978).

Index

Agriculture   209, 210, 211, 212, 213, 214, 215, 216
  area   211, 213, 214, 215
  energy need   220, 221, 222
  production   215, 216, 231, 232, 233
Albedo   140, 141, 142
Amino acids
  in food   207, 211, 212
  in meteorites   67
  on the Moon   68
  primordial synthesis of   83
  terrestrial   116
Ammonia
  cosmic abundance of   60, 62
  in interstellar gas   62, 63, 64
  in terrestrial atmosphere   71, 82
Animal draft as energy source   154
Annihilation of particles   10
  energy   17, 177, 178
Anti-Big Bang (see Big Crunch)
Anti-hadrons   8, 177, 178
Anti-particle   8, 31
Asteroids   69
Atomic nuclei
  binding energy of   14
  magic numbers   13
  relationships   7
Atmosphere
  composition of   108, 109, 110, 117, 118
  climatic impact of   144
  cycling of   127
  energy flux on   145, 146
  evolution of   78, 79, 82, 256
  mass of   107

Background radiation   23, 26, 63, 97
Baryon number, conservation of   5, 106

Beryllium-8   49, 50
Beta-decay   12, 15, 17
Big Bang   20
  and evolution of Universe   20, 28, 136, 250
  and nucleosynthesis   39
Big Crunch   28, 97, 250
Binding energy of atomic nuclei   14, 15
Biomass, dry   194, 195, 199
Biosphere
  animals   194, 208, 209
  biomass, dry   194, 195
  chemical composition of   198, 199
  continental   194, 195, 196, 197, 198
  discoveries in   202, 204
  energy flux on   192, 193, 197, 199, 200, 205
  evolution of   98, 99, 100, 101, 102, 201, 256
  and extinctions   202, 203
  and food source   156, 206, 208, 209
  and forests   114, 195, 198
  future of   260, 261
  impact of on Earth   74, 144
  marine   194, 195, 196, 197, 198
  mass of   107, 193, 194, 197
  photosynthesis in (see Photosynthesis)
  plants in   194, 204, 209
  productivity of   196, 197, 198, 205
  water vapourisation in   204, 205
Black hole
  at centres of galaxies   38, 39, 46
  as energy source   177, 178, 179
  particle production by   10
Bohr, N.   246
Boltzmann, L.   57, 192
  constant   3
Bosons   7, 9
Brain evolution   103

Breeder reactor   171, 172, 188
Bricks   8, 9

Catastrophic events   250, 251, 252, 253, 254, 255, 259, 260 (*see also* Supernova. Galaxis, Comets, Meteorites, Sun, Red Giant, Nuclear War, Cretaceous Period)
Carbon
  as coal (*see* Coal)
  cosmic abundance of   24, 25, 60
  cycling   110, 113, 114
  dioxide   82, 114, 115 (*see also* Greenhouse effect)
  in interstellar gas   62, 63
  as life carrier   89, 90, 91, 94, 95, 96
  monoxide   62, 82, 114, 116
  nucleosynthesis of   49, 50
  terrestrial abundance of   75, 82, 110
  waste (*see* Waste)
Cesium-137   185
Chemical
  bonds and life   89, 90
  energy   16, 17
  evolution   58
  forces   58
Chemicals   131
Clouds of life (*see* Interstellar clouds)
Climate
  future   258, 259
  past   143, 144, 147
  present   142, 146, 147
  regulation of   238, 242, 243
Coal   154, 155
  origin of   160
  resources   160, 161
  uranium energy equivalent of   172
  waste   180, 181
Cold era of Universe   21, 22
Comets
  chemical composition of   65, 66
  structure of   65, 66
  terrestrial impact of   80, 81, 203, 251
  "Tunguska"   66, 257
Conservation law   4
  baryon number   5, 11, 12
  lepton number   5
  weak force   12
Continents
  climatic impact of   144
  drift of   74, 148, 149
  properties of   74
Copernicus' principle of mediocrity   4, 86

Cosmic abundance of elements   24, 25, 60
Cosmic colonies   261, 262, 268, 269, 270, 271
Cosmic rays   55, 56, 203
Creation of particles   10
Cretaceous Period   148, 202, 203, 251, 257
Cryosphere   121 (*see also* Hydrosphere)

Darwin, Ch.   57
Deuterium
  burning of, in stars   46, 47
  as fusion fuel   174, 175, 176, 177
  resources of   161
  synthesis   33, 35
Doppler effect   21
Drake limit   268, 269
Dust, atmospheric   119, 120

Earth
  Archean phase of   72
  atmosphere of (*see* Atmosphere)
  biosphere of (*see* Biosphere)
  catastrophic events on (*see* Catastrophic events)
  chemical composition   75, 125, 126
  chemical evolution on   72, 78
  climate on (*see* Climate)
  continental plates shift on   78
  core of   71, 72, 75
  cosmic environment of   252, 253, 254
  Cretaceous Period of   148, 202, 203, 251, 257
  crust of   72, 73, 77, 125
  geothermal sources of   78
  gravitational field of   106
  hydrosphere of (*see* Hydrosphere)
  lithosphere of (*see* Lithosphere)
  mantle of   72, 73
  properties (mass, radius, etc.) of   73
  proto-   71, 72
  Proterozoic phase of   72
  spaceship   233
  spheres of, five   106, 107
  stability of   256, 257
  surface of   213, 214, 230, 231, 244
  technosphere of (*see* Technosphere)
Economics of energy production   188, 189, 190
Economy principle   4
Ecosphere   234
Einstein, A.   1, 18, 246

Index

Electric charge 3, 248
Electromagnetic
  energy 16
  force(s)
    attractive and repulsive 8, 88
    as carrier of life 88
    relationships between 6
Electron
  capture 35
  conservation law 11, 106, 248
  properties 3, 10
  relationships 7
Electronegativity 58, 59
Electron volt conversion factor 3
Elementary
  forces (*see* Forces)
  particles (*see* Particles)
Elements
  cosmic abundance of 24, 25, 60, 90
  electronegative 58, 59
  heavy 54
  as life carrier 89, 90
  lifetimes of 54
  magic numbers of 13
  periodic table of 14
  superheavy 54
Emden, R. 135
Encephalisation coefficient 103
Energy
  annihilation 15
  beta-decay 16
  black hole's 16
  chemical 16
  conversion factors 3, 138
  of electromagnetic force 16, 136, 137
  fission 16, 136, 137
  free
    civilisation flux of 154, 155, 156, 237, 238, 239, 240, 241
    hydrosphere flux of 122
    ordered 16, 135
    solar, flux 137, 138, 139
    terrestrial, flux 80, 81, 137, 138, 139
  fusion 16, 136, 137
  gravitational 16, 136, 137, 247, 248
  –mass relationship 16
  need for man 152, 153
    past 154, 155
  sink 130, 179, 186, 187, 241, 242
  of strong force 16, 136, 137
  in synthesis of nuclei 16, 136
  thermal 16, 136
  of weak force 16, 136, 137

Entropy
  and life 86, 87
  and matter 130, 132
  and space organisation 231
Environment, human (*see* Radioactivity)
Erosion 125
Eutrophication 115
Expansion of Universe (*see* Universe)
Extraterrestrial
  colonisation 268, 269, 270, 271, 272
  information exchange 265, 266, 267
  intelligence 262, 263, 264, 265, 267

Fatty acids, in food 207
Fermions 7, 9
Fine structure constant 3
Fireball 22, 36
Fish harvest 218, 219
Fission of atomic nuclei
  energy from 16, 17
  fuel for 161, 169, 170, 171, 172, 173
  fuel waste of 180, 181, 182, 183, 184, 185
  and nuclear energy (*see* Nuclear energy)
  terrestrial 136
Fluorocarbons 118, 119
Food
  animal 127
  consumption 206
  and energy use 210, 220, 221, 222, 238
  ocean production 216, 217, 218, 219
  production (*see* Agriculture)
  quality 211, 212
  single-cell protein 222, 223
  vegetable 127
  wood 157
Forces
  electroweak (*see* Unified force)
  chemical 6
  electromagnetic 6, 8, 22, 32, 136, 249
  grand unified 6, 22, 31
  gravitational 6, 8, 22, 31, 106, 136, 247, 248, 249
  nuclear 6, 136
  strong 6, 8, 22, 31, 136, 249
  superunified 6, 22, 30
  unified 6, 22, 31, 32
  weak 6, 8, 22, 32, 136, 249
Forests 114, 195, 198
Fossil fuel (*see also* Coal, Gas, Oil)
  resources 161
  waste 180, 181

Index

Fusion of atomic nuclei
  energy from   17
  and solar energy   136
  synthesis of deuterium in   35
  and thermonuclear power   174, 175, 175, 177, 240, 241

Galaxies (*see also* Milky Way)
  evolution of   37
  explosions in   251, 252
  giant elliptical   38
  irregular   38
  luminosity of   37
  properties of   38
  proto-   37
  quasars   37, 38
  radio-   37, 38
  spiral   38
  stability of   250, 251
  and the Universe   20, 250
Gas
  as an energy source   155
  resources   161
Gas-dust clouds (*see* Interstellar gas)
Genetic information   101, 102
Geothermal
  energy   150, 151
  power   168, 169
Glaciations   147, 148
Gold, T.   225
Grand unified force   6, 11, 22
Gravitation(al)
  attraction   8, 88
  collapse (*see* Black hole)
  decoupling   22
  energy   16
  and pair production   11
  relationships   6, 8, 247, 248, 249
  terrestrial   106
Gravitons   7, 17
Gray, definition   181, 251
Great Extinction   148
Greenhouse effect   141, 144
Gross National Product (GNP)   212

Hadron
  epoch of Universe   21, 22
  stability   12
Heliostats   164, 165
Helium
  flash   44
  recombination   36
  synthesis   34, 35
  in stars   41
  in terrestrial atmosphere   71
  three-alpha-process of   48, 49
Hertz conversion factor   3
Hertzsprung–Russel diagram   40
Homo erectus   102
Homo habilis   102
Homo sapiens   102
Hot era of Universe   21, 22
Hoyle, F.   18, 246
Hubble, constant of   27, 28
  time   28
Human labour   152, 154, 189, 190
Hydro-energy   154, 155, 158, 159
Hydrogen
  binding energy of   15
  burning in Sun   42, 43, 47
  and CNO-process   48
  cosmic abundance of   24, 25, 60
  and fusion power   174
  as life carrier   89, 90, 91, 94, 95
  pp-process of   48
  recombination   36
  synthesis of   32, 33
  terrestrial abundance of   75, 82
Hydrosphere
  climatic impact of   144
  and cryosphere   121, 146
  cycling rate   120, 121, 127
  energy flux on   122, 146
  evolution   78, 79, 256, 257
  mass of   107
  ocean as   121, 122, 216, 230
Hydroxyl   60, 113

Inertial confiment   175
Infrared radiation   141
Intelligence, extraterrestrial   262, 263, 264, 265, 266, 267
Interstellar
  clouds   62, 63
  molecules   63, 64
Iron
  binding energy   15
  "catastrophe"   71
  cosmic abundance of   24, 25, 59, 60
  electronegativity of   59
  nucleosynthesis of   50, 52
  oxidation of   110
  oceanic   218
  terrestrial abundance of   75, 76

Joule conversion factor   3

Index

Kelp 157
Koino-hadrons 8
Koino-particles 31

Laws of Nature 3, 226, 247, 248, 249
Lehrer, T. 105
Lem, S. 245
Lepton 5, 8, 21, 22
Life
  chemical composition of 94, 95, 96
  cosmic 85, 93, 94, 95
  definition of 85, 86, 95
  electromagnetic force and 88
  elementary particles and 88
  energy and 88, 92, 93
  evolution of 97, 98, 99, 100
    genetic 101
  photolysis, water and 100
  self-regulation of 87, 95
  self-reproduction of 87, 95
  spontaneous origin of 87, 95
  terrestrial 85
Light, velocity of 3, 247, 248
Lithosphere
  cycling rate of 127
  chemical composition of 75, 125, 126
  evolution of 77, 79, 256
  mass of 107
Lukewarm era of Universe 21, 22

Magic numbers
  electronic 13
  nuclear 13, 15, 52, 54
Magnetic confinement 175
Main Sequence stars 40, 47
Man (*see also* Homo)
  evolution of 102, 260
  food consumption 206
  future of 260, 261, 262
  genetic information 102, 260
Marcus Aurelius 105
Materials
  cycling rate of 109, 126, 127, 233, 234, 235
  energy use of 131, 236, 237
Matter 4, 9, 105
Mediocrity principle 4
Metals
  energy use of 131
  ores of 129, 235
Meteorite(s)
  Allende 67
  amino acids in 67
  chemical composition of 66, 67
  impact on Earth of 72, 80, 81, 203, 252
  Murchison 67
  origin of 66
Methane
  cosmic abundance of 61
  in interstellar gas 62, 63, 64
  resources 161
  in terrestrial atmosphere 71, 82
Milankovitch hypothesis 142, 143, 144, 145
Milky Way 20, 37, 38
  colonisation of 272, 273
  habitable planets in 263, 264
  structure of 38
Minerals, number of 126
Mole scale 2, 3
Moon
  amino acids on 67, 68
  properties of 69
  tidal energy of 151
Mortar 8, 9

Neutrino
  clumping 37
  decoupling 22
  generation, three 9
  e-neutrino 7, 9
  Lepton epoch 22, 34
  mu-neutrino 7, 9
  tau-neutrino 7, 9
Neutron
  decay 34
  mass 3
  nucleosynthesis 52, 53
  properties 12
  star
    evolution of 13, 43, 45
    synthesis 32
Newton, I. 30
Newton's principle of universality 4
Nitrogen
  atmospheric 110, 116
  binding energy of 15
  cosmic abundance of 24, 25, 59, 60
  cycling rate of 116
  electronegativity of 59
  fertiliser 116, 117
  as life carrier 89, 90, 94, 95, 96
  lifetime of 54
  nucleosynthesis 48, 49

Nitrogen (*cont.*)
  oceanic   218
  terrestrial abundance of   75, 76
Noble gases   13, 14
Non-solar energy   150, 151
Nuclear
  decay and geothermal energy   150
  fission power   155, 169, 170, 171, 172, 182, 183, 240, 241
  fusion power   174, 175, 176, 177, 185, 186, 241
  power and breeder reactors   171, 172
  resources   161, 172, 176
  utilisation   169, 170, 171, 172
  war   119, 259, 260
  winter   119
Nucleons
  stability of   12
  binding energy of   15
Nucleosynthesis
  Big Bang   30, 31, 32, 33, 34, 35
  CNO-process   48
  pp-process   48
  r-process   52, 53, 54
  s-process   52, 53, 54
  stellar   47, 48
  three-alpha-process   48, 49
  X-process   56
Nuclides magic number   15

Ocean   121 (*see also* Hydrosphere)
Oceanic thermal power   158, 159
Ockham, W.   1
  principle of economy of   4
  "razor"   19
Oil, as energy source   154, 155
  resources   161
Ores   128, 129
Otec   158, 159
Oxygen
  atmospheric   110, 111
  binding energy of   15
  burning, for technology   112
  cosmic abundance of   24, 25, 59, 60
  cycling rate   111
  electronegativity of   58, 59
  isotopes on Earth   74
  as life carrier   89, 90, 91, 94, 95, 96
  and oxidation   110
  nucleosynthesis   50
  sphere   76
  terrestrial abundance of   75, 76
Ozone   112, 113, 119

Pair production   10
Particle
  boson   7, 9
  brick   8, 9
  creation of   9
  fermion   7, 9
  elementary forces of   8, 9
  gluon   7, 9
  graviton   7, 9, 17
  hadron   7 (*see also* Hadrons)
  lepton   7 (*see also* Leptons)
  as life carriers   88
  mortar   8, 9
  muon   7, 9
  neutrino   7 (*see also* Neutrinos)
  photon   7, 9
  quark   7 (*see also* Quarks)
  stability of   11, 88
  tau   7, 9
  W-boson   7, 9
  super-   7, 30
Peat, resources   161
Periodic table   14
Phosphorus
  binding energy of   15
  cosmic abundance of   24, 25, 59
  electronegativity of   59
  as life carrier   94, 95, 96
  lifetime of   54
  oceanic   218
  terrestrial abundance of   75
Photon(s)
  cosmic composition of   25, 26
  conservation law of   9
  creation and annihilation of   10
  decoupling of   22
  epoch of Universe   21, 22, 36
  and pair production   10
Photosynthesis   111, 112, 114, 160, 200, 201, 205
Planckian era   30
Planck's constant   3, 248
  units of   2
Planetary systems   70
Planet(s)
  accretion of   71
  Earth   68, 69 (*see also* Earth)
  habitable   263, 264
  Jupiter   69, 254, 255
  Mars   69
  material stability of   106
  Mercury   68
  Neptune   69
  origin   70

Pluto 69
properties of 68, 69
proto- 70
Saturn 69
Uranus 69
Venus 68
Plants (see Biosphere)
Plutonium 171, 172
Poisons 131, 132
Polymers 84, 85
polynucleotide 83
polypeptide 83
polysugar 83
Population world 226, 227, 228, 229, 270
Positrons, and helium synthesis 46
Primordial heat 78
Primordial soup 98, 99
Principle of Nature 4
Prohibition laws 5
Proteins, food 207, 208, 218, 219
Protobionts 97, 98
Proto-Earth 71, 72
Protogalaxies 37
Proton
anti- 11
and beta-plus decay 34
decay 11, 12, 248
koino- 11
quarks 8
properties 3, 12
synthesis 32
Protoplanets 71
Protostar 39, 41

Quark(s)
anti- 8, 31
and Big Bang 20
bonding forces 6
bottom 9
charmed 9
down 9, 32
elementary forces and 8
epoch of Universe 21, 22
generation, of three 8, 9
and hadrons 8, 32
koino- 8, 31
and nucleons 32
relationships 7
strange 9
soup 33
top 9
up 9, 32
Quality of life 243, 244

Rainfall 121, 204
Radioactivity
lethal dose of 181, 182, 251
natural 80, 182, 184
Radioactive waste 171, 177, 180, 181, 182, 183, 184, 185
Radio waves 62, 63, 265, 266
Reactor fission (see Nuclear fission power)
Recycling 131
Red giants (see also Sun)
evolution of 40
helium burning in 49
properties of 43
Red shift 21
Doppler effect and 21
Rees, M. 225
Rindler, W. 18

Salam, A. 1
Schrödinger, E. 57
Silicon
cosmic abundance of 59, 60, 61
oxide 61, 62
Single-cell protein 222, 223
Solar
cells 164, 165
constant 139, 141, 144
energy flux on the Earth
and agriculture 211
future 238, 239, 240
past 79, 142, 143, 144, 201
present 136, 137, 138, 139, 140
spectrum of 140, 141
extraterrestrial power station 166, 167
flares 251, 253
heating 163, 164, 165
orbital power stations 166, 167
plantations 156, 157
photochemical power 164, 165
photovoltaic power 164, 165
power stations 164, 165
system (see Planets)
wind 65
Stars
black dwarf 40, 44, 55
blue giant 40, 41, 44
components of 42
era of Universe 22, 37
evolution of 37, 44, 55
gravitation of 40
luminosity of 40
Main Sequence 40, 41, 42, 44, 55

Stars (*cont.*)
  multiplicity of   41
  neutron   43, 44, 45, 55
  origin of   37
  and planetary nebulae   40
  proto-   39, 40
  red giant   40, 41, 43, 44, 55, 254, 255
  red supergiant   40
  supernova   44, 45, 55, 167, 203
  white dwarf   40, 43, 44, 55
Stratton, R.   225
Strong force
  as binding force   13
  decoupling of   22
  energy of   16
  and nucleosynthesis   46
  relationships   6
  in Universe's evolution   32
Strontium-90   185
Sulphur
  binding energy of   15
  coal burning and   62, 163
  cosmic abundance of   24, 25, 59
  dioxide   118
  in hydrogen sulphide   61, 82
  electronegativity of   59
  lifetime of   54
  oxidation of   110
  terrestrial abundance of   75, 76, 82
Sun
  chromosphere of   42
  companion of   254
  evolution of   42, 43, 44
  hydrogen burning in   47
  as life carrier   93
  luminosity of   40, 42
  as Main Sequence star   40
  properties of   42
  red giant   254, 255
  spots   42
  rotation of in Galaxies   39, 42, 253, 254
  wind from   65
Supergravity   30
Supernova
  explosion of   44, 45, 251
  terrestrial impact of   80, 81, 203, 251, 252, 253
  as uranium source   167
Superparticle   7, 30
Superunified force
  range of   5
  relationships of   6

and temperature   5
very hot era of Universe and   22, 30

Technosphere
  climatic impact of   144
  cycling rate of   127, 128
  mass of   107
Thermal waste of   186, 187 (*see also* Energy sink)
Thermodynamics   86, 179
Thermonuclear
  bombs   175
  reactor   174, 175, 176, 177, 241 (*see also* Nuclear fusion)
  waste   185, 186
Thorium as nuclear fuel   161, 162, 167, 168
Tidal energy   151, 157, 158
Time scale   2
Tritium
  synthesis   34, 175
  reaction   35, 175
  as fusion fuel   174, 175, 176, 177
Tunguska (*see* Comets)

Ultraviolet radiation   62, 92, 98, 112, 141
Unified force relationships   6, 22
Universality principle   4
Universe
  age of   28
  and Big Bang   (*see* Big Bang)
  and Big Crunch (*see* Big Crunch)
  cold era   32, 33
  composition of   21, 23, 24, 25
  density   21, 249, 250
  elements, abundance of, in   24, 25
  energy flux in   137, 138
  entropy of   136
  eras of, five   21, 22
  expansion of   20, 21, 27, 28, 249, 250
  and Fireball   22, 36, 37
  future of   28, 247, 248, 249, 250
  hadron epoch of   22, 33
  homogeneity of   27
  horizon of   19, 23, 26
  hot era of   22, 30, 31
  isotropy of   27
  lepton epoch of   22, 33, 34
  life carriers in   96, 97, 98
  lukewarm era of   22, 31
  mass of   24

Index   289

neutrinos in 21, 24
order of 136
Planckian era of 30
photon epoch of 22, 36
proton stability in 11
radius of 28
schematic representation of 19
star era of 22, 37
temperature of 22, 23, 249
transparency of 23
unified force era of 22, 31, 32
very cold era of 22, 32, 37
very hot era of 22, 30
volume of 26
wholeness of 19
Uranium
  abundance of 173
  binding energy of 15
  breeding of 170, 171, 172
  vs. coal 162
  cosmic abundance of 24, 25, 59, 60
  electronegativity of 59
  as fuel 169, 170, 171, 172, 173
  as geothermal source 78
  lifetime of 54, 167, 170
  nucleosynthesis 51, 52, 55
  resources 161, 173
  -235 52, 55, 167, 170, 171, 172, 173
  -238 52, 55, 167, 170, 171, 172, 173
  terrestrial abundance of 75, 167

Vegetable material cycling 127
Velocity of light 3, 5
Very hot era of Universe 21, 22, 30
Vitamines 207, 211
Virus 102
Volcanoes
  activity of 74, 257
  dust emission from 120
  Krakatoa 257

Waste
  energy 179, 180, 181, 186, 187, 241, 242
  material 132, 133
  radioactive 171, 172, 180, 181, 182, 183, 184, 185
Water (see also Hydrosphere)
  cosmic abundance of 60, 92
  cycling rate of 121, 122, 123, 124, 125
  dissociation of 92
  as life carrier 91, 92
  and photolysis 92, 100
  and power production 158, 159
  properties of 91
  quality of 122, 123, 124
  structure of 91
  volcanoes and 82
Waves, energy of 139, 158
Weak force
  decoupling 22
  energy of 16
  and nucleosynthesis 46, 175, 176
  and number of nuclides 15
  and particle stability 12
  relationships 6
Weinberg, A.M. 105, 135
Wheeler, J.A. 246
White dwarf 40, 43, 44, 45
Wind energy 139, 154, 155
  and power production 157, 158
Wood
  as energy source 154, 155, 156
  food source 157
  silviculture 156
World ship 271, 272

X-particles 20, 31
X-process 56

/530T222E>C1/

DATE DUE

DEC 03 1986			

Demco, Inc. 38-293